GREEN BUILDING:
A PROFESSIONAL'S GUIDE TO
CONCEPTS, CODES, AND INNOVATION
Includes IgCC Provisions

Anthony C. Floyd, AIA, LEED AP and Allan Bilka, R.A.

DELMAR
CENGAGE Learning

Australia • Brazil • Japan • Korea • Mexico • Singapore • Spain • United Kingdom • United States

Green Building: A Professional's Guide to Concepts, Codes, and Innovation
Anthony C. Floyd and Allan Bilka

Delmar Cengage Learning Staff:

Vice President, Technology and Trades Professional Business Unit:
Gregory L. Clayton

Director of Building Trades:
Taryn Zlatin McKenzie

Executive Editor: Robert Person

Acquisitions Editor: Helen Albert

Product Manager: Vanessa Myers

Director of Marketing: Beth A. Lutz

Senior Marketing Manager: Marissa Maiella

Marketing Coordinator: Rachael Torres

Senior Production Director: Wendy Troeger

Production Manager: Sherondra Thedford

Senior Content Project Manager:
Stacey Lamodi

Senior Art Director: Benjamin Gleeksman

ICC Staff:

Senior Vice President, Business and Product Development: Mark A. Johnson

Deputy Senior Vice President, Business and Product Development: Hamid Naderi

Technical Director, Product Development:
Doug Thornburg

Director, Project and Special Sales:
Suzane Nunes Holten

Senior Marketing Specialist:
Dianna Hallmark

For product information and technology assistance, contact us at
**Cengage Learning Customer &
Sales Support, 1-800-354-9706**

For permission to use material from this text or product,
submit all requests online at **www.cengage.com/permissions.**
Further permissions questions can be e-mailed to
permissionrequest@cengage.com

Library of Congress Control Number: 2011933043

ISBN-13: 978-1-111-03511-2
ISBN-10: 1-111-03511-3

ICC World Headquarters
500 New Jersey Avenue, NW
6th Floor
Washington, D.C. 20001-2070
Telephone: 1-888-ICC-SAFE (422-7233)
Website: http://www.iccsafe.org

Delmar
5 Maxwell Drive
Clifton Park, NY 12065-2919
USA

Cengage Learning is a leading provider of customized learning solutions with office locations around the globe, including Singapore, the United Kingdom, Australia, Mexico, Brazil and Japan. Locate your local office at: **international.cengage.com/region**

Cengage Learning products are represented in Canada by Nelson Education, Ltd.
Visit us at www.InformationDestination.com. To learn more about our green product line, visit www.TheGreenDestination.com.
For more learning solutions, Please visit our corporate website at **www.cengage.com**

Printed in the United States of America
1 2 3 4 5 6 7 14 13 12 11

CONTENTS

ACKNOWLEDGEMENTS

We would like to thank Hamid Naderi, PE, Deputy Senior Vice President of International Code Council Product Development, who identified the need for this type of publication. He also served as the project manager for this publication through a rapidly code changing environment. We would also like to thank Steve Thorsell, ICC-ES Director of Sustainability Evaluation Programs for his valuable input. We thank Vanessa Myers and the staff of Delmar Cengage Learning for their editorial and publication expertise. Finally, we would like to thank all those who contributed along the way towards the development of the 2012 International Green Construction Code.

Anthony C. Floyd, AIA, LEED AP and Allan Bilka, R.A.

September 2011

ABOUT THE INTERNATIONAL CODE COUNCIL®

The International Code Council® (ICC®) is a nonprofit membership association dedicated to protecting the health, safety, and welfare of people by creating better buildings and safer communities. The mission of ICC is to provide the highest quality codes, standards, products and services for all concerned with the safety and performance of the built environment. ICC is the publisher of the family of the International Codes® (I-Codes®), a single set of comprehensive and coordinated model codes. This unified approach to building codes enhances safety, efficiency and affordability in the construction of buildings. The Code Council is also dedicated to innovation, sustainability and energy efficiency. Code Council subsidiary ICC Evaluation Service issues Evaluation Reports for innovative products and reports of Sustainable Attributes Verification and Evaluation (SAVE).

Headquarters: 500 New Jersey Avenue, NW, 6th Floor, Washington, DC 20001-2070
District Offices: Birmingham, AL; Chicago. IL; Los Angeles, CA
1-888-422-7233
www.iccsafe.org

PREFACE

Look deep into nature and then you will understand everything better.
Albert Einstein

Green building and sustainable construction is rapidly on its way towards being the norm. What began not long ago as a fringe building movement has become standard policy for many governmental entities. Several state and local government entities have adopted green building criteria based on the International Green Construction Code (IgCC), or allow the IgCC as an alternate or option. Such are the CALGreen in California and various IgCC based programs in Florida, North Carolina, Oregon, Maryland, Phoenix, Scottsdale and several other municipalities. From the U.S. federal agencies, state agencies, counties, townships, municipalities and globally, many countries and cities, there are several hundred established policies and mandates for public facilities to be LEED green certified at various levels, certified with the Green Globes or have some level of compliance with SBTool.

Sustainable construction requires whole building systems thinking in the design, construction, inspections, verification/commissioning and operations phases. It requires a collaborative design and integrated project management process and also requires a rigorous inspection, verification and building system commissioning process.

Is green building ready for prime time? Is green building warranted in light of higher thresholds that have been set in current energy, mechanical, plumbing and building codes? Are green codes a natural progression of public health, safety and general welfare? Will green building codes and standards accelerate or hinder the principles behind green building and sustainable communities?

What about the obligations of design professionals, builders, developers, realtors, appraisers and financial institutions? How about the responsibilities of owners in the operation and maintenance of buildings? What about the level of responsibility of building regulatory authorities and their ability to enforce requirements after the certificate of occupancy is issued?

The answers to these questions and many others, including the coverage of provisions of the IgCC (edition available at the heels of the historic final public hearing for the 2012 IgCC, the first model green building code, in Phoenix, Arizona), and some rating systems will unfold in the following pages. We will find that a fundamental understanding of design intent, rating tools, professional standards of care, contractual arrangements and construction codes and standards is required in order to answer these questions and fully comprehend their implications. In addition, all stakeholders, including the public and industry, must be willing to face reality, understand the dire potential consequences of ignoring our environmental responsibilities, and embrace change.

Beyond these questions, what is certain is that new opportunities for the design and building industry will be driven by outcome-based design, accountable construction, commissioning and the on-going operation of and demand for green buildings. New and expanded services will be needed for energy modeling, plan review, inspections, building commissioning, building performance monitoring and post-occupancy and outcome-based evaluations. Green building's success, however, is dependent on an informed, capable and willing building industry. This book provides a wide array of valuable information that is intended to empower the building industry and enable it to move further down the path toward the critical goal of producing a sustainable built environment.

The International Code Council (ICC) and ASHRAE both have made historical steps to codify green building into an overlay code and standard. With the creation of the ASHRAE 189.1 Standard Project Committee and later the ICC Sustainable Building Advisory Committee (SBTC), green building criteria were codified within the regulatory framework of minimum baseline requirements with the support of the American Institute of Architects, ASTM International, Illuminating Engineering Society, US Green Building Council, and other building construction associations. In response, this book has morphed into a document that provides a backdrop and valuable insight into the history, development, relevance, implementation and environmental benefits of numerous fundamental sustainability related concepts and practices. Many of these concepts and practices have lead to the creation of, and now form the basis of, newly developed green building rating systems, standards and codes such as the 2012 International Green Construction Code. This book also gives insight into the relationship of these green tools to each other and to conventional building codes such as the International Building Code (IBC). We hope that code officials, design professionals, contractors, manufacturer's and policy makers will find this information useful in their quest to produce a built environment which is safe and sustainable.

ABOUT THE AUTHORS

Anthony Floyd is a registered architect and Senior Green Building Consultant for the City of Scottsdale. During the 1980s, Anthony served as a building inspector, plans examiner and later as the building official for the City of Scottsdale from 1988 -1995. In 1995-96, he participated in a 9-month International Honors Program (IHP) focusing on sustainability, development, and global ecology in the countries of England, India, Philippines, New Zealand, and Mexico. After returning to Scottsdale, Anthony helped to establish Arizona's first Green Building Program in 1998. He maintains the city's regionally based green building program criteria, conduct public outreach and is facilitating the transition to the city's adopted International Green Construction Code (IgCC). He served on the drafting committee for both the IgCC and the National Green Building Standard and chairs the site sustainability working group on the ASHRAE 189.1 Standards committee for the Design of High-Performance Green Buildings. Anthony has a civil engineering and architecture degree from Penn State University and a Master's degree in public administration from Arizona State University. He is frequently called upon to lecture on regenerative building design principles and has taught classes on green building rating systems and building codes at Arizona State University.

Allan Bilka is a registered architect, serving as Senior Staff Architect with the International Code Council (ICC). His primary responsibilities at ICC are related to green and sustainable building, where he serves as Secretariat in the development of the International Green Building Code and ICC Staff Liaison in the development of the ICC 700 National Green Building Standard. Allan is also involved in the development of training and certification materials to support the IgCC, serves as an instructor and speaker on sustainable topics, and has written various articles, workbooks and white papers related to green and sustainable building, primarily on ICC's behalf. Prior to coming to ICC, he practiced architecture for 15 years in the Northwest Indiana and Chicago, Illinois areas, designed and built his home using many green and sustainable concepts in 1984, and studied architectural design at Ball State University in Muncie, Indiana, from 1970 to 1972. It was at Ball State that he was introduced to environmental studies and "alternative" building concepts and strategies, including many which are now considered fundamental to green and sustainable building.

AN INTRODUCTION TO GREEN BUILDING

We do not inherit the earth from our ancestors; we borrow it from our children.
—Native American proverb

A NEW AGE OF CONSTRUCTION AND ECOLOGICAL IMPACT

Once upon a time, we could cast off our waste and Mother Nature would naturally clean up after us. Buildings were constructed of natural materials, there were far fewer buildings, there weren't so many of us, and natural ecosystems were typically able to cope with our environmental impacts. A building constructed of natural materials could be easily reclaimed by the earth. In fact, most buildings were naturally "green" and "sustainable" and humans almost instinctively lived sustainable lifestyles. They didn't have any other choice and they didn't know any other way. Even burning wood or coal, prior to the industrial age, had relatively few negative implications for the environment. You could drink the water of most rivers and lakes without worrying that it might pose a threat to your health.

Technological advancements and the population explosion have changed all that. We've created new materials that natural ecosystems are often not equipped to deal with. And, in addition to the population explosion, each individual's environmental or ecological footprint has exploded as well. On average, each human consumes significantly more resources and produces significantly more waste than ever before. We are already stepping on each other's ecological footprint, and the earth simply isn't large enough to sustain a scenario where all humans on the planet live a lifestyle similar to that which we currently enjoy in the United States. Even "natural" waste by-products are now created in such massive quantities that ecosystems often cannot absorb them all. Yet all too many of us unwittingly create undue pressures on our environment, basing many of our decisions solely on our economic ability to do so and our seemingly natural urge for more of everything. We tend to accumulate, use, and discard things simply because we can afford to do so, without regard for the environmental consequences of our actions. These habits are rooted in our cultural past, a past that in many respects is no longer relevant to today's technology-driven environmental scenario. However, we are in the midst of a new awareness of environmental consequences as we face the limits of our natural resources, technological solutions, and the earth's ability to absorb the negative by-products of human activities.

The media reports green building related news with such frequency that it has become impossible for any single individual to absorb it all. Governmental entities at local, state, and national levels are increasingly adopting green building mandates for publicly funded facilities. The adoption of green benchmark building codes and standards is taking place at the local and state level as both voluntary and mandatory programs. Some states, such as California, have adopted mandatory green building codes for both public buildings and private-sector commercial and residential buildings. In many parts of the country, local governmental jurisdictions have developed voluntary green building programs that offer incentives, including expedited plan review, reduced permit fees, zoning bonuses, and assistance with documentation, to facilitate the approval of alternative materials and methods. Some jurisdictions have green building requirements for special zoning districts, such as downtown and high intensity areas.

Even in regions where local governmental jurisdictions have not implemented green building programs, independent voluntary programs continue to appear with ever-increasing frequency. Green building has become a powerful marketing tool for manufacturers, builder/contractors, developers, design professionals, and building owners.

THE IMPACT OF BUILDINGS

In the United States, buildings account for:

- 40% of municipal solid waste[1]
- 30% of raw material use[2]
- 12% of potable water use[3]
- 49% of all energy produced[4]
- 77% of electricity produced[5]
- 46.9% of carbon dioxide (CO_2) emissions[6]

In the not too distant past, traditional thinking told us that the transportation sector was the major consumer of electricity and contributor to greenhouse gases, such as CO_2. Most people are surprised that the building sector consumes more energy than any other sector. Most of this energy is produced from burning fossil fuels, making the building and operation of buildings the largest emitter of greenhouse gases on the planet. According to the U.S. Energy Information Administration (EIA), nearly half (46.9%) of all CO_2 emissions in 2009 came from the building sector. By comparison, transportation accounted for 33.5% of CO_2 emissions, and industry, just 19.6% (see Figure 1-1).

---

[1] U.S. EPA Office of Solid Waste, *Municipal Solid Waste in the Unites States: Facts and Figures* (2009).

[2] David M. Roodman and Nicholas Lenssen, Worldwatch Paper 124, "A Building Revolution: How Ecology and Health Concerns Are Transforming Construction" (Worldwatch Institute, Washington, DC, 1995).

[3] U.S. Geological Survey (2004).

[4] U.S. Energy Information Administration (2008).

[5] Ibid.

[6] Ibid.

Even if one does not subscribe to the concepts of global warming and climate change, it is difficult to deny the fact that buildings consume tremendous amounts of energy, the vast majority of which is produced by power generation plants that rely on greatly depleted natural resources: fossil fuels. Indeed, energy shortages and reliance on foreign sources of fossil fuels have become natural security issues, and buildings play a large part in this scenario (see Figure 1-2).

The waste mountains that are ever more frequently appearing on our horizons are also due, in large part, to building construction, including the activities that buildings house and the habits of the building occupants with regard to waste

Figure 1-1

United States CO$_2$ emissions by sector.

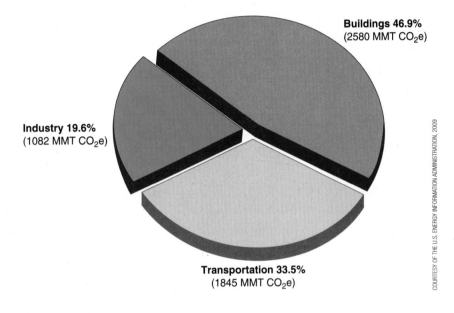

Buildings 46.9%
(2580 MMT CO$_2$e)

Industry 19.6%
(1082 MMT CO$_2$e)

Transportation 33.5%
(1845 MMT CO$_2$e)

COURTESY OF THE U.S. ENERGY INFORMATION ADMINISTRATION, 2009

Figure 1-2

United States energy consumption by sector.

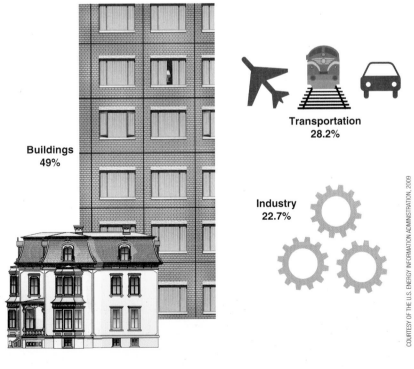

Transportation
28.2%

Buildings
49%

Industry
22.7%

COURTESY OF THE U.S. ENERGY INFORMATION ADMINISTRATION, 2009

Defining Green

Sustainable Design

Buildings and the built environment play a major role in the human impact on the natural environment and on the quality of life.

- Sustainable design integrates consideration of resource and energy efficiency, healthy buildings and materials, ecologically and socially sensitive land use, and an aesthetic sensitivity that inspires, affirms, and ennobles.

- Sustainable design can significantly reduce adverse human impacts on the natural environment while simultaneously improving quality of life and economic well-being.

Source: "Declaration of Interdependence for a Sustainable Future," World Congress of Architects, Chicago, June 1993.

disposal. Material shortages are becoming more commonplace, farmland is competing with urban sprawl, biomass energy sources are competing with food supply, and higher food costs and shortages are becoming more prevalent, particularly in underdeveloped countries. So many things seem to have come full circle or, more accurately, reached their practical limits. And the design, construction, and operation of buildings are major factors in all of these scenarios.

Stunning technological advances during and since the industrial revolution have only made it easier for us to consume more natural resources, particularly during the era of cheap energy. Until about 100 years ago, thermal comfort and lighting were mostly achieved with passive building design. Heating was achieved by a compact design and a fireplace or stove, cooling by opening windows to the wind and shading them from the sun, and lighting by windows.

Even as technology advanced, as long as cheap energy was available, it became easier and more convenient to rely on inefficient heating/cooling equipment and artificial lighting during daylight hours. Heating/cooling systems are often needlessly oversized and passive design features are not included or considered in the building design. Though more energy is used worldwide to heat water than for any other human-related purpose, storage tank water heaters, which are an inefficient way to provide hot water (because energy is used to keep the water hot at all times regardless of demand, producing significant standby loses), are still the norm. Hot water distribution lines are not designed to efficiently deliver the hot water to the point of use. There is usually a significant loss of heat energy, and water is often wasted while waiting for hot water. In the production and delivery of electricity from fossil fuel, about 70% of the original energy is lost (see Figure 1-3). Most manufactured components and products used in buildings and their systems have not been designed with efficiency and the environment in mind.

Many nations, states, and municipalities are struggling to cope with these realities and their possible future dire implications. Some have vowed to take action and do something about them. For example, the

Figure 1-3

Approximately 70% of the original energy is lost when a fossil fuel is converted to electricity.

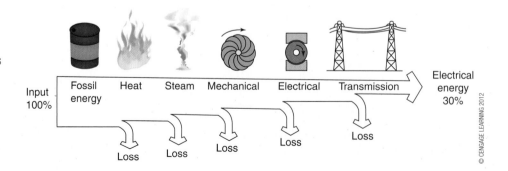

© CENGAGE LEARNING 2012

Defining Green

Sustainable Development

1. An extension of the view of the built environment beyond shelter, to include "energy harvesting, waste management and reuse, food production and distribution, water harvesting and handling, as well as facilities for recreation, health, education, commerce, etc."
2. Reduction of construction processes that damage the environment in favor of those that restore it
3. The strict implementation of reuse and recycling of building materials
4. Encouraging the creation of self-reliant communities to reduce transportation, energy, and material use
5. A return to well-established methods of design that conserve energy and natural resources
6. A further examination and exploration of the potential of self-help in the "making, remaking and use" of sustainable settlements
7. The encouragement of community participation in the design and construction process
8. Urban energy; harvesting, forestry, food production, and hydrology; and wildlife management supported by the involvement of United Nations (UN) agencies

Source: AIA Environmental Resource Guide, 1992.

U.S. Conference of Mayors, which is composed of the mayors of U.S. cities with populations of 30,000 or more, unanimously signed an agreement in 2006 to meet the targets of the 2030 Challenge.[7] Basically, the challenge predicts dire consequences if our use of fossil fuels for energy production continues along its current path, and therefore challenges us to significantly decrease that use by specific amounts until buildings become carbon neutral (or net zero, using no more energy than they consume) by 2030 (see further discussion in Chapter 3). In accordance with the resolution, the U.S. Conference of Mayors is working to create the fossil-fuel reduction standards for all new buildings. In order to meet these targets, as well as many other environmental goals and initiatives, many jurisdictions have turned to green, sustainable, and high-performance building policies for both public- and private-sector projects.

DEFINING GREEN BUILDING

Generally speaking, green building is building with a conscious effort to reduce the negative impact that buildings have on the natural environment. However, building green does *not* guarantee that a structure will have absolutely no negative impact on the environment. In fact, buildings that meet only the criteria for the lowest tier of some green building programs in the past performed only marginally

---

[7]Resolution adopted by the members of the U.S. Conference of Mayors at the 74th Annual Conference, Las Vegas, Nevada, 2006.

better, if at all, from an environmental perspective, when compared to buildings of conventional code-compliant construction.

At the opposite end of the scale, the highest-performing green buildings are sometimes considered high-performance buildings. High-performance buildings are generally thought of as buildings that perform well beyond minimum code requirements or minimum green building criteria incorporating many of the best practices that current technology can reasonably provide. Just as there are high-performance equipment, appliances, and automobiles, there are also high-performance buildings. Such buildings are optimized at a whole-building-system scale in terms of energy efficiency, indoor environmental quality, and resource utilization.

The concept of sustainability, as related to the environment, takes the goals of environmental harmony to levels well beyond those of green building. The goal of sustainability is that present generations enjoy a lifestyle and use technology and resources in a controlled manner that ensures that the same quality of life and resources will be available to all future generations. Theoretically, buildings that are sustainable have no negative environmental impact. However, none of the green or "sustainable" rating systems, programs, or standards currently available inherently produces buildings that are truly sustainable in this ideal sense, though they are slowly moving toward that goal.

Green building can be thought of as a first step toward sustainable building practices. Sustainability, on the other hand, is an ideal, almost unattainable goal, much like that of producing safe buildings in which lives are never lost. Building codes approach their ideal goals by means of features such as fire sprinkler protection and fire-resistance-rated noncombustible construction. Although these measures have saved many lives, they have not totally prevented loss of life. Similarly, if we are limited to the natural resources of planet earth, and we hope to sustain human existence indefinitely, does that mean that we should not be allowed to use any natural resources unless they are renewable? If that is too extreme, what percentage of the earth's available resources should be available to each generation? The highest-performing green buildings use the best that current technologies will economically and reasonably allow to come as close as possible to the idealistic goals of sustainable building, but there are still many difficult questions that must be answered.

As stated earlier, prior to the industrial age, most buildings were naturally sustainable, as virtually all materials were natural, were easily returned to the earth, and/or were used in low quantities, ensuring that they had far less environmental impact and were available to future generations. Now, short of returning to those past ways of life, in order to be sustainable, we must create environmentally responsible technologies, construction methods, and infrastructure to manage the environmental impact of all harmful technologies and methods of construction. In addition, we must discover new sources of natural resources, and move toward the use of renewable resources wherever they are practical.

The benefits of building green speak directly to these concerns and beyond, addressing a triple bottom line that supports environmental, economic, and community well-being in the areas of site development, water, energy, material resources, indoor environmental quality and building operations and maintenance (see Table 1-1).

Table 1-1
Benefits of Sustainable Design

	Environmental	Economic	Societal
Siting	Land preservation, reduced resource use, protection of ecological resources, soil and water conservation, reduced energy use, less air pollution	Reduced costs for site preparation, parking lots, and roads; lower energy costs due to optimal building orientation; less landscape maintenance cost	Improved site use and increased transportation options for employees
Water efficiency	Lower potable water use and pollution discharges to waterways; less strain on aquatic ecosystems in water-scarce areas; preservation of water resources for wildlife and agriculture	Lower first cost (for some fixtures); reduced annual water costs; lower municipal costs for wastewater treatment	Preservation of water resources for future generations and for agricultural and recreational uses; fewer wastewater treatment plants and associated impacts
Energy efficiency	Lower electricity and fossil-fuel use, less air pollution, and fewer carbon dioxide emissions; lowered impacts from fossil-fuel production and distribution	Lower first costs, when systems can be downsized due to integrated energy solutions; lower fuel and electricity costs; reduced peak power demand; reduced demand for new energy infrastructure; lowering energy costs to consumers	Improved comfort conditions for occupants; fewer new power plants, transmission lines, and associated impacts
Materials and resources	Reduced use of virgin resources, healthier forests due to better management practices, lower energy use for material transportation; reduced strain on landfills, increase in local recycling markets	Decreased first costs due to material reuse and use of recycled materials; lower waste disposal costs; reduced replacement costs for durable materials; reduced need and costs for new landfills	Fewer landfills and associated impacts, greater markets for environmentally preferable products, decreased traffic due to use of local/regional materials
Indoor environmental quality	Better indoor air quality, including reduced emissions of volatile organic compounds (VOCs), carbon dioxide, and carbon monoxide	Organizational productivity improvements due to improved worker performance, lower absenteeism, reduced staff turnover, lower disability/health insurance costs, and reduced litigation	Reduced adverse health impacts; improved occupant comfort, satisfaction, and employee productivity
Commissioning: operations and maintenance	Lower energy consumption, reduced air pollution and other emissions.	Lower energy and maintenance costs; reduced costs associated with occupant/owner complaints; improved building durability and longer equipment lifetimes.	Improved occupant productivity, satisfaction, health, and safety.

ABOUT THIS BOOK

Green Building: A Professional's Guide to Concepts, Codes, and Innovation is designed for practitioners and students preparing to be a part of the design and building industry. The exemplary green building projects of today are rapidly becoming tomorrow's standard. The industry is transforming itself through a combination of market forces, technology advancements, product innovation, governmental policies, standards, and codes. One hundred years ago, insurance companies and the public health movement laid the foundation for modern building and health codes as we know them today. Within the last 10 to 30 years, energy efficiency has become a vital component of building codes. Today we see the scope of codes and standards expanding to encompass the core principles of green building in the light of public health, safety, and general welfare. By means of newly developed green codes and standards, the building industry will be provided with an effective tool to incorporate green building systems and practices as part of the building permit process. The future of construction is the normalization of green.

This book is divided into three major parts. Chapter 2 presents the context of the natural and built environments, including a discussion of natural ecosystems, environmentally responsive building, green terminology, and the economics of green. Chapter 3 discusses a multitude of governmental and market-based programs that have gained momentum over the last couple of decades. These programs and initiatives have increasingly been adopted by governing jurisdictions and supported by the building industry as a means to save energy, reduce waste, and lower operating costs, while simultaneously developing a market niche that produces a multitude of societal benefits.

Chapters 4 through 10 address seven major environmental impact areas of building, from planning and construction to building operations and maintenance. Each chapter includes a discussion of applicable green building tools, codes, and standards. Chapter 4 discusses the broader implication of planning communities in terms of land use, density, transportation options, pedestrian connectivity, and neighborhood development patterns. Chapter 5 addresses the specifics of site development, including stormwater management, heat-island mitigation, landscape design, and preservation of natural resources. Chapters 6 through 9 get into the specifics of water efficiency, material resources, energy efficiency, and indoor environmental quality. Finally, Chapter 10 discusses the vital importance of building commissioning, maintenance, and operations that are the final lynchpin in the process toward truly achieving green buildings and communities.

THE EVOLUTION OF AN ANCIENT IDEA

2

This chapter provides some environmental background to help understand the relationship between the natural world and the human-built.

THE DISRUPTION OF ECOSYSTEM SERVICES

How much of the earth's ecological balance can we disrupt before we pass a threshold in the ability of nature to maintain life-support services? We should now know enough to understand that services produced by nature's ecosystems are essential for human life. These services include provision of clean water and flood control, healthy soil, pollination of crops, providing habitats for fisheries, and countless other benefits that support human life and well-being.[1]

Human society has never had a more pivotal time to understand its relationship with and dependence on nature. From time immemorial we have undervalued some of the most basic resources on which we depend, including the air we breathe, the water we drink, and the ability of the earth to support a wide variety of life. The cumulative impact of human activity on the natural systems that produce these resources and support services, particularly over the past 100 years, makes it impossible for us to take them for granted any longer.

WE DON'T PROTECT WHAT WE DON'T VALUE

Over countless generations, people have tended to place relatively little emphasis on protecting the ecosystems, in great part because we have neither understood nor appreciated their value. If we each dumped all of our trash and garbage in our own landfill in our backyards, personally extracted coal for electric power generating plants, or cleared countless forests for development, we might have a better appreciation for the impact of our daily actions. But we do not. We are removed from those direct experiences, and out-of-sight becomes out-of-mind. Few seldom consider the implications of those actions. Without a firm understanding of the value of nature's services to the quality of our lives and to our ability to impact and destroy or maintain life, we are unlikely to make the necessary compromises needed to protect them.

---

[1]Daily, Gretchen C., *Nature's Services: Societal Dependence on Natural Ecosystems* (Washington, DC: Island Press, 1997).

All of the world's population depends on nature's life-support services, not simply for the goods they provide but also because of the critical role they play in maintaining the global atmosphere, regulating the earth's weather patterns, filtering much of the waste products produced by society, preserving watersheds, controlling soil erosion, and preventing floods and droughts. The value of these ecosystems far exceeds that of the marketable resources we commonly associate with them.

On a global scale, different groups of people are now living at one another's expense, as is apparent in the disruption and overexploitation of the earth's resources and the inability of the earth to safely absorb waste. Whereas the levels of environmental impacts were once small, local, and reversible, they have now reached global proportions and if ignored carry irreversible consequences.

HUMANS AND ECOSYSTEMS

An ecosystem is the set of organisms living in an area, their physical environment, and the interactions between each other and their environment. There is often no clear distinction between "natural" and "human-dominated" ecosystems.

Ecosystem services are the conditions and processes through which natural ecosystems and the species that make them up, sustain human life. They maintain biodiversity and the population of ecosystem goods, such as seafood, forage, timber, biomass fuels, natural fiber, and many pharmaceuticals, industrial products, and their precursors. In addition to the production of goods, ecosystem services are the actual life-support functions, such as cleaning, recycling, and waste removal.[2]

Ecosystems are ancient, the product of billions of years of evolution, and have existed in forms very similar to those seen today for at least hundreds of millions of years. They are absolutely pervasive, but unnoticed by most humans going about their daily lives. Noticed or not, human beings depend completely on the continuation of natural cycles for their very existence.[2] If the life cycles of predators that naturally control most potential pests of crops were interrupted, it is unlikely that pesticides could satisfactorily take their place. If the carbon cycle were badly disrupted, rapid climatic change could threaten the existence of life as we know it.

For millennia, humanity has drawn benefits from these cycles without causing global impact. Yet, today, human influences can be detected in the most remote reaches of the planet: deep below earth's surface in ancient aquifers, far out to sea on tiny tropical islands, and up in the cold, thin air high above Antarctica.[2] Virtually no place remains untouched, chemically, physically, or biologically, by the actions of humankind.

Defining Green

Ecosystem Services

Conditions and processes through which natural ecosystems, and the species that make up, sustain and fulfill human life. They are actual life-support functions, including purification of air and water, detoxification and decomposition of wastes, generation and renewal of soil and soil fertility, pollination of crops and natural vegetation, maintenance of biodiversity, protection from the sun's harmful ultraviolet rays, partial stabilization of climate, and moderation of temperature extremes.

Source: Daily, Gretchen C., *Nature's Services: Societal Dependence on Natural Ecosystems* (Washington, DC: Island Press, 1997).

---

[2]Daily, Gretchen C., *Nature's Services: Societal Dependence on Natural Ecosystems* (Washington, DC: Island Press, 1997).

Human history cannot be understood in a vacuum. All human societies have been, and still are, dependent on complex and interrelated physical, chemical, and biological processes. These include the energy produced by the sun, the circulation of the elements crucial for life, the geophysical processes that have caused the continental landmasses to migrate across the face of the globe, and the factors regulating climatic change. These constitute the essential foundations for the way in which the various types of plants and animals, including humans, form complex interdependent communities.[3]

Ecosystems are not static. It takes thousands of years to move from bare rock through lichens and mosses to ferns, plants, and eventually trees to create a climax forest that can survive for long periods. As ecosystems develop and change, so do the plants and animals that can be supported unless a major disruption occurs caused by either human interference or a natural disaster such as an earthquake, volcano, or cataclysmic storm.

To fully understand the individual parts of an ecosystem, it is necessary to see them as part of a bigger picture. All the parts of an ecosystem are interconnected through a complex set of self-regulating cycles, feedback loops, and linkages between different parts of the food chain. If one part of an ecosystem is disrupted, there will be an adverse reaction elsewhere in the system.

Over the last 10,000 years, human activities have brought about major changes in the world's ecosystems. The expansion of settlements, the creation of fields and pastures for agriculture, the clearing of forests, and the draining of marshes and wetlands have all reduced the habitats of almost every type of plant and animal.[3] In addition, humans have transplanted plants and animals around the world, often with adverse consequences.

One of the greatest changes in human history has been the unprecedented and rapid increase in population over the last 250 years. The total number of people in the world first reached 1 billion in about 1800, and it had taken about 2 million years to reach this level (see Figure 2-1). The next billion was added in about 130 years. A further billion took about 30 years from 1930 to 1960. The next billion was added in only 14 years (by 1974), and the next billion took about 13 years (until 1987). In the next 12 years another billion was added, bringing

Figure 2-1

World population growth of modern humans.

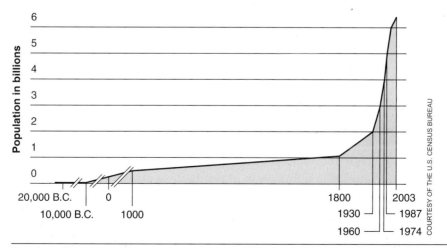

[3]Ponting, Clive, *A New Green History of the World* (New York: Penguin Books, 2007).

the world's population to 6 billion in 1999. The number of people in the world is projected to reach 7 billion by 2012.

This unprecedented rise in population and resource consumption has had a profound effect on the environment. All of these people have had to be housed and therefore the number and size of human settlements across the globe has increased dramatically. Increased consumption of the earth's resources have led to increased pollution and ecosystem disruption around world.

ENERGY AND POLLUTION

Until the nineteenth century human societies had limited availability and access to energy resources. The last two centuries have been characterized not just by a vast increase in energy consumption but by an increasing use of and reliance on nonrenewable fossil fuels (coal, oil, and natural gas).[4]

All previous human societies had depended upon renewable sources of energy—humans, animals, water, wind, and wood. The fact that they had usually mined forests with little, if any, thought given to conservation or replanting meant that this energy crisis was self-inflicted—the result of a short-sighted approach repeated century after century. Only when the shortage became acute did societies have no alternative but to exploit coal on a large scale, even though it was an inferior fuel. It was the start of the switch to dependence on nonrenewable sources of energy. The first major exploitation of the world's fossil-fuel reserves, created from the great tropical forests that existed over 200 million years earlier, began in the seventeenth century.[4] It marked a fundamental discontinuity in human history—a move from the energy shortage that had characterized human history up until this point to societies that depended upon rapidly growing and very high energy use.

One of the most significant energy developments of the last two centuries has been the use of fossil fuels to provide a highly convenient from of secondary energy—electricity. The major increase in electricity production and consumption took place in the twentieth century and relied upon a number of linked developments: (1) the construction of ever larger power stations; (2) the construction of high-powered transmission lines and grids to distribute power locally and nationally.[4] The overwhelming majority of electricity has always been generated from fossil-fuel-fired power stations—at first coal and then oil and natural gas.

Once the electricity generation and distribution infrastructure was in place, the highly convenient form of energy used in factories and homes for lighting, heating, and power marked a fundamental shift in energy patterns. For the first time energy was easily available at the flick of a switch, and this was one of the main driving forces behind greater energy consumption. Electricity made possible far greater automation of production processes through the use of machine tools and provided the energy for completely new industries that required significant amounts of electricity. It also provided the basis for the growth of domestic energy

[4]Ponting, Clive, *A New Green History of the World* (New York: Penguin Books, 2007).

use and the industries that make the consumer electrical appliances that consume the domestic energy. The United States was the first country to be electrified on a major scale, in the urban areas in the 1920s and a decade later in the countryside. Nearly every household in the industrial world is now connected to the electric utility grid that provides power for lighting, heating, and cooking, as well as power for a vast array of household goods, such as refrigerators, freezers, televisions, washing machines, dishwashers, and computers.[5]

Over 85% of the world's energy now comes from nonrenewable fossil fuels (40% from oil, 25% from coal, and 21% from natural gas; see Figure 2-2). The transition to fossil fuels has been accomplished by a spectacular rise in consumption. The world's annual consumption of coal is now about 500 times greater than it was in 1800. The annual consumption of oil is now about 380 times greater than it was a century ago. The consumption of natural gas rose 175-fold in the twentieth century.[5] For most of the last two centuries, fossil fuel has been cheap. In fact, the price of oil actually fell for long periods in the twentieth century.

There are increasing signs that the world is entering a new stage in its energy history characterized by the beginning of the end of the great oil-based energy boom (see Figure 2-3). The decline in the importance of oil is likely to take place over a considerable period of time. Production will peak long before reserves become exhausted—the most likely date is estimated around 2020, followed by a long decline in output. The decline in natural gas output is projected to occur in conjunction with the fall in oil production.

Although the countries of the industrialized world are now more energy efficient than they were a century ago, this has not stopped a significant increase in energy consumption. Further increase in energy efficiency through new technologies will not stop rising demand for energy and higher levels of energy consumption. Ultimately, the impact of energy consumption on the environment comes from the limited availability of resources, ecosystem disruption during resource extraction and the pollution generated in the process of producing the energy.[5]

Figure 2-2

World energy use.

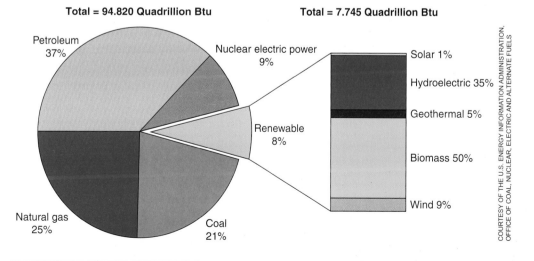

Total = 94.820 Quadrillion Btu   Total = 7.745 Quadrillion Btu

Petroleum 37%
Nuclear electric power 9%
Solar 1%
Hydroelectric 35%
Geothermal 5%
Renewable 8%
Biomass 50%
Wind 9%
Natural gas 25%
Coal 21%

COURTESY OF THE U.S. ENERGY INFORMATION ADMINISTRATION, OFFICE OF COAL, NUCLEAR, ELECTRIC AND ALTERNATE FUELS

[5]Ponting, Clive, *A New Green History of the World* (New York: Penguin Books, 2007).

Figure 2-3

Annual oil production.

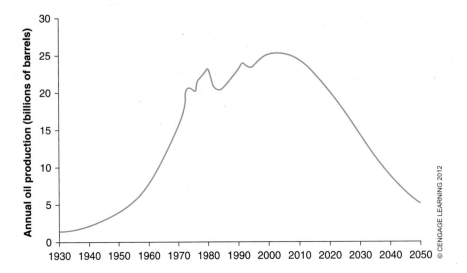

Pollution has been one of the distinguishing features of every human society. For much of human history the chief struggle was over sanitation and obtaining clean water supplies. As populations grew and more people lived in cities, this problem became ever more critical. The development of industrial societies introduced new pollutants on a major scale and brought about new risks to human health and greater damage to the environment. Pollution was at first mainly localized and confined to an area of a city, river, mine, or waste dump. But as industrialization increased, the affected areas grew to entire regions, continents, and, ultimately, oceans. By the late twentieth century pollution became a clear threat to the global systems that make life possible on earth.[6]

The earth is a closed system. Everything created on the earth stays on the earth. Disposal of industrial or consumer-product waste merely means placing it somewhere else on the planet. Our understanding of the consequences of waste and pollution has always lagged behind the creation of pollutants and waste. Our response to controlling pollution and waste is vital to sustaining human life. Rachel Carson, in her pivotal book *Silent Spring* (1962), revealed the large-scale proliferation of harmful chemicals and their impact on humans and ecosystems.

The course of human history over the last two centuries has progressed at a pace never experienced, along with a series of interlinked issues as a result of population growth, industrialization, resource consumption, and energy use. It is the interaction of all of these changes that has produced the current state of our environment (see Table 2-1).

The problem for all human societies has been to find a way of extracting the necessary food, clothing, shelter, and other goods from the environment in a way that does not render the environment unsustainable or incapable of supporting human life. The environment is somewhat resilient and can withstand some degradation within the boundaries and limits of a given ecosystem. The continuing challenge for humans is to know the limits of ecosystems and find the political, economic,

[6]Ponting, Clive, *A New Green History of the World* (New York: Penguin Books, 2007).

Table 2-1
The World in the Twentieth Century

	Increase 1900–2000
World population	× 3.8
World urban population	× 12.8
World industrial output	× 35
World energy use	× 12.5
World oil production	× 300
World water use	× 9
World irrigated area	× 6.8
World fertilizer use	× 342
World fish catch	× 65
World organic chemical production	× 1,000
World car ownership	× 7,750
Carbon dioxide in atmosphere	+ 30%

© CENGAGE LEARNING 2012

and social means to respond accordingly. Some societies have succeeded in finding the right balance; others have failed.[7]

A MATTER OF SCALE

Humans have altered nature for 4 million years. However, this is the first time in human history that we have altered ecosystems with such intensity, on such scale and with such speed. To see just how peculiar the last 100 years was, it helps to adopt long perspectives of the our human past.[8]

Before the industrial revolution began, we had at our disposal the muscle power of our bodies and of domesticated animals, the power of wind and water, and the chemical energy stored in wood and other biomass materials. No other century in human history can compare with the twentieth century for its growth in energy use. We have consumed more energy since 1900 than in all of human history before 1900.

The growth of cities marked a turning point in human and environmental history. Cities had for many centuries dominated civilizations, but in the twenty first century they became home for the majority of people on the planet. Urban growth has had significant effects on water-, land-, and energy-consumption patterns. Urban areas take in water, food, and oxygen and discard sewage, garbage, and carbon dioxide.

Urban roofs and roads prevent water from percolating into the earth and thereby increases surface runoff. A large city like Chicago, built on sodden prairie, changed the hydrology of surrounding waterways to accommodate growth and

---

[7]Ponting, Clive, *A New Green History of the World* (New York: Penguin Books, 2007).

[8]McNeill, J. R., *Something New Under the Sun* (NewYork: Norton, 2000).

development. Growing cities also need timber, steel, cement, brick, plumbing and building systems, food, and fuel. Prior to the age of rail, most of this came from within the region. Besides shipping over water, railroads and trucks made it possible for a significant share of goods and materials to come from distant regions and lands, dispersing the environmental effects across the country, and thus enlarging the environmental footprint of cities.

Many of the environmental buffers such as clean water supply and unused accessible resources that helped societies weather difficult times in the past are now gone. The general policy of the last one hundred years was to try to make the most of available resources, make nature perform to the utmost, and hope for the best. We have increasingly brought about self-imposed environmental constraints in the form of the planet's capacity to absorb the wastes and to sustain overall impacts of our human activities. Although ecological devastation has happened in the past at a local scale, this is the first time human societies will be constrained at a global scale.[9]

BUILDINGS AND THE ENVIRONMENT

Our early ancestors were inextricably tied to their environment, because their survival depended on it. They used the resources available to them to create shelter, to hunt or gather food, and later to farm and travel.[10] They built houses to keep out the elements—rain, wind, sun, and snow—and created an environment favorable to their comfort and well-being.

Before the advent of the industrial era, humans depended on natural sources of energy and available local materials in forming their habitats according to their physical needs. Over many centuries, people everywhere appear to have learned to work in concert with their climate, available resources, and environmental constraints. Climate and local resources have shaped the rhythm of their lives in the form and function of habitat. In the warm humid environments, local inhabitants often live in huts with loosely woven walls that allow the slightest breeze to pass through, while people who live in the desert often construct houses with thick mass walls and small openings to keep out heat and the glare of the sun.[11]

Humans have demonstrated throughout history an ability to adapt their built environment to local climatic and geographic conditions. We have the wind towers of the Middle East designed to capture prevailing winds to cool internal spaces; shade canopies to protect desert city street markets; mushrabeyeh wooden screens to shade and ventilate second story urban bay windows in Cairo; and sliding shoji screens to control ventilation and allow for privacy in traditional Japanese homes. As these examples illustrate, each solution is a balanced response to climate, local resources, technology, and culture.[12]

[9]McNeill, J. R., *Something New Under the Sun* (New York: Norton, 2000).

[10]Keeler, Marian, and Bill Burke, *Integrated Design for Sustainable Building* (Hoboken: Wiley, 2009).

[11]Fathy, Hassan, *Natural Energy and Vernacular Architecture* (Chicago: University of Chicago Press, 1986).

[12]Jones, David Lloyd, *Architecture and the Environment* (Woodstock: Overlook Press, 1998).

Historically, there was little option but to use local resources sparingly and to maximum benefit. The processing of early building materials was limited in scale. In many regions around the world, earthen bricks were usually dried in the sun, and the husks and stalks of harvested crops were also used to fire them. Limestone and chalk were baked for cement and gypsum plaster. Iron and other metals were employed in limited quantities. This situation changed in the nineteenth century, with wide-scale industrialization bringing techniques that greatly improved the performance of construction materials. Industrialization often meant overextraction of local resources, pollution, waste, and ecological devastation.

DESIGN WITH CLIMATE

The Native Americans of North America in various regions possessed a remarkable ability to adapt their traditional building forms to their particular environmental settings. An awareness of climate was integrated with innate craftsmanship to solve problems of comfort and protection. The results were low-impact and regionally responsive buildings based on appropriate technology.[13]

Native Americans evolved in a broad variety of climatic environments, from the cold-cool Northern territories to the warm-hot areas of the South, from the dry Southwest areas to the humid parts of the Southeast (see Figure 2-4).

The tribes entering the cold zone encountered extreme cold and relatively scarce fuel. Under these circumstances, the conservation of heat became essential, so

Figure 2-4

Climate zones of North America.

© CENGAGE LEARNING 2012

[13]Olgyay, Victor, *Design with Climate* (Princeton: Princeton University Press, 1963).

Figure 2-5

Plan and section view of an igloo deflects wind and insulates with snow.

their shelters were compact, with a minimum of surface exposure. The Eskimo igloo is a well-known solution to the problem of survival in extreme cold. The low hemispherical shelters deflect the winds and take advantage of the insulating value of the snow that surrounds them. The smooth ice lining that forms on their interior surface is an effective seal against air seepage, and their entry tunnels are oriented away from the prevailing winds to reduce drafts and prevent the escape of warm air (see Figure 2-5). The heat retention of this type of structure makes it possible to maintain a temperature of 60°F inside when the outside temperature is −50°F. Such structures may be heated by a small lamp supplemented by body heat.

The temperate climatic regions, offering a naturally favorable environment, made fewer thermal demands on its inhabitants. There is a corresponding diversity and flexibility in the structures of these peoples. The dwelling choice of the Plains Indians was the tipi, a conical structure of poles covered by skin, which effectively shed wind and rain and was easily heated from a central fire, as shown in Figure 2-6.

In contrast, the hot-arid zone made extreme demands on the constructors of tribal dwellings. Characterized by excessive heat and glare, this area required that shelter be designed to reduce heat impacts and provide shade. The southwestern tribes often built communal structures for mutual protection. Structures such as those of the Pueblos were constructed of massive adobe walls and roofs, which have good insulative value and the capacity to delay heat impacts for long hours, thus reducing the daily heat peaks. They also used very small windows to reduce heat gain and glare. By clustering the buildings together, the amount of exposed surface was reduced. Pueblo structures were usually placed on an east-west axis, thereby reducing morning and afternoon heat impacts on the two end walls in the summer and receiving a maximum amount of southern sun during the winter months (see Figure 2-7).

The hot-humid regions presented two major challenges: the avoidance of excessive solar radiation and the evaporation of moisture by breezes. As a solution, the

Figure 2-6

The conical structure of a tipi provide shelter, and heat comes from a central fire.

© ISTOCKPHOTO/WILLIAM WALSH

Figure 2-7

Pueblo structures are positioned to reduce morning and afternoon heat.

© ISTOCKPHOTO/S. GREG PANOSIAN

native inhabitants built their villages to allow free air movement, and the scattered individual units were mixed into the shade of surrounding flora. The Seminoles, for example, raised large gable roofs covered with grass to insulate against the sun and provided large areas of shade over the dwellings, which had no walls. The steep angle and extensive overhang of these roofs protected against rainfall, and the floors were elevated to keep them dry and to allow air circulation underneath.

There is a marked correlation between climate zones and the locations in which roof types commonly occur. Flat and vaulted roofs have historically appeared the hot-arid regions, and pitched roofs are found in the wet-temperate climates. Both domes and vaults are popular in hot-arid regions with clear skies and where large timbers are scarce. The radiation of high sun positions is diluted on the round dome surface. This results in lower surface temperatures, which are further reduced by wind.[14]

It is more than coincidence that groups of different cultures on different continents appear to have similar solutions to common environmental conditions. These solutions work with, not against, the forces of nature to create low-impact, comfortable, and healthy living environments. Such buildings may be called environmentally responsible, ecologically sound, sustainable, or green.

DEFINING GREEN BUILDING

Green building has been defined by many people and in many different ways. Some emphasize environmentally compatible and low-impact building materials. Others emphasize energy performance and the use of renewable energy, such as solar. Still others emphasize indoor air quality and healthy interior material finishes that minimize or eliminate potentially harmful substances, such as formaldehyde, volatile organic compounds (VOCs), or even electromagnetic fields (EMFs).

Since the 1990's, a number of regionally based green building programs were established that provided useful definitions. In Scottsdale, Arizona, green building is defined as "a whole systems approach utilizing design and building techniques to minimize environmental impact and reduce the energy consumption of a building while contributing to the health of the building's occupants"—in other words, building a healthy, resource- and energy-efficient building. It's a matter of making the right design choices by choosing the right materials, resources, and methods of construction. In the *International Green Construction Code* (IgCC), green building is not defined in the traditional sense, it is defined by building in accordance with the code. The intent of the code, in addition to safeguarding public health, safety and general welfare, is to safeguard the environment and to reduce the negative potential impacts and increase the positive potential impacts of the built environment on the natural environment.

Green building isn't necessarily about using new materials and technologies, or incorporating pre-industrial materials such as adobe, heavy timber, or low-tech solutions. The crux of green building is an integrated and whole-systems approach to building. In other words, doing more with less and considering the building holistically—as the sum of many parts which intrinsically impact each other. This approach tends to minimize environmental impacts associated with energy, water, material resources, and indoor environmental quality. The fact is that there are levels or shades of green, just as there are levels of quality, durability, health, and safety. Only the highest levels of green building approach true sustainability.

As witnessed by the current state of world affairs, we live in an increasingly populated and interconnected world of limited resources. No longer are buildings isolated islands. As long as energy, water, and material resources are needed to build and sustain our buildings, we will need to be smarter about the way we build

---

[14]Olgyay, Victor, *Design with Climate* (Princeton: Princeton University Press, 1963).

and the way we use resources. We no longer live in a world of bountiful resources with unlimited capacity and endless landfills for waste.

From a global environmental, health, and natural resources perspective, we need green buildings more than ever. In the not-too-distant future, green building will no longer be referred to as green. It'll just be good building sense.

HIGH-PERFORMANCE, WHOLE-BUILDING DESIGN

The term *high-performance building* is sometimes interchangeably used with *green building*. According to the U.S. Office of Energy Efficiency and Renewable Energy (EERE), a high-performance commercial building "uses whole-building design to achieve energy, economic, and environmental performance that is substantially better than standard practice." This requires that the design team fully collaborate from the project's inception in a process often referred to as integrated design.

Whole-building or integrated design considers site, energy, materials, indoor air quality, acoustics, and natural resources, as well as their interrelation with one another. In this process, typically a collaborative team of architects, engineers, building users, owners, and specialists in indoor environmental quality, materials, and energy and water efficiency utilizes systems thinking to consider building systems holistically, examining how they best work together to save energy and reduce the environmental impact.[15]

INTEGRATED DESIGN

Integrated design is a holistic design approach that is vital to green buildings. It requires that the various design conditions and variables be considered as a unified whole in determining the desired outcome of the project.

Because every design decision produces a cascade of multiple effects, rather than an isolated impact, successful integrated design requires a necessary understanding of the interrelationship of each material, system, and spatial element. It requires all the project team member to think holistically about the project rather than focus solely on an individual part or component of the project.[16]

In an integrative design process, the project team might include the owner, architect, engineering consultants (civil, structural, mechanical, electrical, fire protection), builder, consultants (energy, daylighting, acoustics, commissioning), building operators, and potential users. Depending on the nature of the project, additional expertise may be needed for specialized systems such as vegetated roofs or on-site renewable energy systems. Integrated design leads to integrated project delivery and involves the builder early in the design process (see Figure 2-8).

Integrated practice usually requires additional design time in the schematic phase in order to allow for communication and coordination between the design team members, including the specialty consultants and contractor. Fundamental sustainability

---

[15]Kibert, Charles, *Sustainable Construction* (Hoboken: Wiley, 2008).

[16]Keeler, Marian, and Bill Burke, *Integrated Design for Sustainable Building* (Hoboken: Wiley, 2009).

Figure 2-8

Integrated project delivery involves the builder early in the design process.

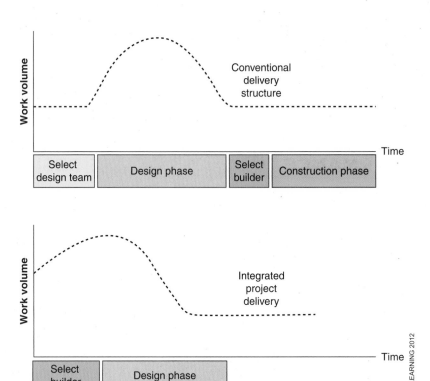

© CENGAGE LEARNING 2012

Defining Green

Green and Ecological Design

Green or ecological design here means building with minimal environmental impacts, and, where possible, building to achieve the opposite effect; this means creating buildings with positive, reparative and productive consequences for the natural environment, while at the same time integrating the built structure with all aspects of the ecological system (ecosystem)…over its entire life cycle.

Source: Yeang, Ken, *The Green Skyscrapter* (Munich: Prestel, 1999).

goals need to be established early in the design to set meaningful targets against which to assess opinions and level of achievement.

It is important to stress that integrated design and project delivery is a methodology. No methodology is perfect, and no project can be completely sustainable. There are always compromises and trade-offs to be made by weighing the merits and complementary effects of various options. That said, there is usually an optimal underlying design concept that leads to a solution that is uniquely connected to project conditions and constraints. Every project presents its unique host of opportunities and challenges.

SUSTAINABLE AND ECOLOGICAL BUILDING

Besides integrated design, other green-building-related terms have been used since the early days of the environmental movement, including *ecological design, environmental design, alternative building, natural, low-impact, environmentally responsible building, earth friendly, and design.* Each term seeks to acknowledge and promote the interrelationship of and interdependence between the natural and built environments.

The progress of green building requires greater understanding and consideration of the environmental and human impact of the built environment, as well as incorporation of natural systems into the design and building process. Without greater understanding of basic ecological

Table 2-2
Conventional versus Ecological Design

Issue	Conventional Design	Ecological Design
Consideration of ecological conditions	Standardized processes followed with regardless of location	Design is integrated with local climate and ecology
Natural habitat	Design controls nature and increases predictability in order to meet defined human/community needs	Heavily relies on natural materials and energy in order to better partner with natural habitat and resources
Knowledge base	Focused on specific design disciplines	Involves many different types of design and sciences
Design criteria	Economics and convenience	Ecosystem and human health and environmental impact
Whole systems	Divides systems differently than natural processes	Strives to produce design that integrates whole systems and natural divisions
Energy source	Usually includes nonrenewable or high-risk resources such as fossil fuels or nuclear power	Renewable resources such as sun, wind, or biomass
Materials	Materials that are not renewable and meet only purpose	Materials that are durable, renewable, multi-purpose, and recyclable
Pollution	Only takes into account legal considerations	Amount and type of waste suited to the ecosystem's ability to absorb it
Ecological and economic	Ecology and economic agendas are seen as opposing	Agendas seen as compatible and long-ranging, integrated approach
Ecological accounting	Limited to compliance with mandatory requirements	Considers impact from start of project to end

© CENGAGE LEARNING 2012

principles in conjunction with building design, construction, and operations, green buildings may cease to evolve beyond merely showcase buildings.[17]

Sim Van Der Ryn and Stuart Cowan, authors of *Ecological Design* (1996), defined ecological design as "any form of design that minimizes environmentally destructive impacts by integrating itself with living processes.[LM2]"[18] Unlike design that destroys landscapes and nature, ecological design seeks solutions that integrate human-created structures with nature in a symbiotic manner (see Table 2-2).

CARBON-NEUTRAL DESIGN

Carbon-neutral design is a term used for building design with the intent to reduce carbon emissions. Carbon dioxide (CO_2) is a major greenhouse gas. Greenhouse gases (GHGs) trap heat below the earth's atmosphere in the same way that glass traps heat from solar radiation in a greenhouse. This trapping of heat increases temperatures and leads to climate change. Buildings are important contributors to CO_2 emissions, primarily as a result of energy

---

[17]Kibert, Charles, *Sustainable Construction* (Hoboken: Wiley, 2008).

[18]Van der Ryn, Sim and Cowan, Stuart, *Ecological Design* (Washington, DC: Island, 1996).

consumption for heating, cooling, and lighting, and are therefore logical targets for mitigation in an attempt to reduce global warming. Carbon-neutral buildings aim to not only reduce energy use but also to consume energy from a carbon-free (non-fossil-fuel generated) source of electricity.[19]

THE ECONOMICS OF GREEN

Under our predominant economic system, the resources of the earth are treated as capital. Trees, wildlife, minerals, water, and soil are simply commodities to be sold or developed. Their price is simply the cost of extracting them and turning them into marketable commodities. This system assigns no value to the benefits and ecological services of nature.[20]

Conventional economics encourages both the producer and the consumer to use up resources at whatever rate current market conditions dictate. It assumes that as one material or source of energy becomes scarce, its price will rise, and this will encourage the development of substitutes.

Markets prices do not reflect true costs. Some commodities, such as air, are treated as free goods. Without environmental regulations, pollution levels will rise because companies have no price mechanism for the pollution emitted into the air or the waste by-products that are discharged into the air, water, or landfills. Society bears the cost of such pollution through diminishing health and negative environmental consequences.

Few economists have tried to deal with these deeper questions concerning economics and the environment. Hazel Henderson has criticized the fragmentation of economic thought and its failure to take into account human dependence on the eco-services provided by the natural world to support life on earth. Herman Daly argues that the demands of economic activities on the ability of ecosystems to regenerate raw material and absorb waste must be "kept at ecological sustainable levels as a condition of sustainable development."[21]

Because few people understand the broader implications of these systems, green initiatives are often expressed in the most limited economic terms, such as the number of jobs created or the profitability of green products. Unless their true environmental costs are recognized in terms we can all understand, we run the risk of sacrificing the long-term survival of nature's services to our short-term economic interests.[22]

GREEN BUILDING COSTS

The days of making the business case for green building and why it is important have passed. Whereas design decisions of the past were primarily driven by simple first-cost payback, today's robust economic analysis tools, such as return on

---

[19]Grondzik, Walter, *Mechanical and Electrical Equipment for Buildings* (Hoboken: Wiley, 2010).

[20]Ponting, Clive, *A New Green History of the World* (New York: Penguin Books, 2007).

[21]Daly, Herman, *Beyond Growth* (Boston: Beacon Press, 1996).

[22]Ponting, Clive, *A New Green History of the World* (New York: Penguin Books, 2007).

investment and life cycle costing, are employed to account for the long-term performance and value of buildings. First or capital costs for constructing a building clearly affect operation and maintenance costs. Generally speaking, the less money invested in capital costs for a quality building, the more costs there will be for long-term operation and maintenance— hence the expression "penny wise and pound foolish".

Defining Green

Life Cycle Cost (LCC)

LCC takes into account all costs of acquiring, operating, maintaining, and disposing of a building or building system.

You get what you pay for. If a green material, product, or system costs more, it is usually because it's more durable and energy efficient, is easily repairable, and requires less maintenance, or it contributes to a healthier and more productive indoor environment. It's not a question of whether one can afford green building, but rather how much one is able to afford. We must keep in mind that environmentally responsible products are not the most expensive products in the building market. Composite decking (recycled plastic/scrap wood), such as shown in Figure 2-9, is highly durable and requires low maintenance, and its price is comparable to that of higher grades of cedar or redwood.

High-efficiency toilets (1.28 gallons per flush [gpf]) are competitive with middle to high-end toilets. There are also common building materials used every day that are considered green, including engineered lumber products (OSB, composite headers and beams), fly ash in concrete, low-e windows, and recycled-content floor tiles that are less expensive than many imported tiles.

All buildings provide some level of environmental accountability, whether they incorporate energy efficiency, water conservation measures, natural light and ventilation, or environmentally benign materials. All buildings fall somewhere along the eco-spectrum.

Green buildings designed to use fewer resource and to support the health of their inhabitants have often been viewed as more expensive to build than conventional buildings. In recent years, this widespread perception has been proven to be wrong. The volatility of energy prices and the long-term trend of rising demand for finite and rapidly depleting fossil fuels make green and energy efficiency a cost-effective risk-reduction strategy. Any additional costs associated with green building can be minimal. From energy savings alone, the average payback time for a green building is 6 years.[23] Additional benefits include reduced water use, infrastructure savings, and health and productivity gains.

It has been proven that green buildings can be cost-effective, whereas conventional development and design can be be risky and financially imprudent. When accounting for capital cost, operation, and maintenance, green buildings can have a substantially lower-cost and lower-risk option than business as usual. Over 20 years, the financial payback commonly exceeds any additional cost by a factor of between four and six.[24]

A 2007 survey by the World Business Council for Sustainable Development found that business leaders believe that green building is, on average, 17% more expensive than conventional design. However, in a 2007 study conducted by Greg

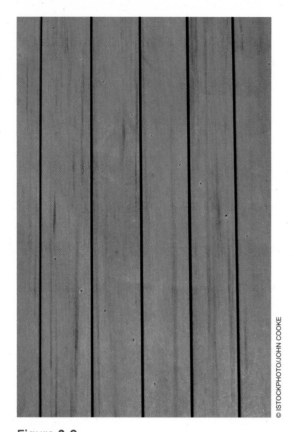

© ISTOCKPHOTO/JOHN COOKE

Figure 2-9

Composite deck materials.

Kats based on LEED certified building (see Chapter 3 for a discussion of LEED), more than three-quarters of the buildings in the data set had green premiums between 0% and 4%. The largest concentration (69 buildings) was between 0% ands 1%. The median cost increase was 1.5%.

In the data set of the study, there are more LEED Platinum-certified green buildings with little or no green premium (0% to 2%) than with a large premium (10% or more), suggesting that the cost premium depends more on the skill and experience of the design and construction team and on the choice of green strategies than on the level of greenness. In addition, early integration of green building objectives into an integrated design and project delivery process can ensure that a cost-effective project is achieved.

Additional up-front soft cost of green building, including energy modeling and building commissioning, is typically offset by savings resulting from improved building performance. Improved insulation and windows can reduce the size of the heating and cooling systems. Increased daylighting can reduce the need for daytime lighting. Optimization of design options is vital to any building design project and is even more critical for green buildings to be cost-effective and achieve long-term benefits. For example, an overabundance of glazing or outside ventilation can impact the ability to heat or cool a building thereby increasing energy costs. Prudent choices and trade-offs must be made in light of up-front costs and long-term benefits. If one is interested in producing a building that uses cutting-edge technologies regardless of cost or benefits, the overall building cost is likely to be prohibitive, and that should be expected.

Buildings are complex systems, consisting of multiple components that are affected by the site conditions, building operations, and by occupant activities and behavior. Differences in site characteristics, climatic conditions, and local regulations may render many green building features impossible to achieve or may make them easily achievable regardless of green objectives. Thus additional costs associated with green building may have more to do with local building practices, community values and development constraints than the fact that a particular building has green features (see Table 2-3).

Table 2-3
Green Cost Factors

Client Goals and Motivation

- Clear objectives and priorities

Location and Site Characteristics

- Topography, hydrology, ecology, zoning, history and access to services
- Building type and use
- Occupant activities/requirements

Regional Environmental Issues and Community Expectations

- Environmental protection, preservation, air quality, open space, traffic mitigation, security, community amenities, property values
- Environmentally responsible development

Local Building Regulations

- Building code amendments, planning, water, and transportation ordinances
- Established development standards and design guidelines
- Locally appointed planning and development review boards

Designer Team and Contractor's Familiarity with Green Design

- Experience with green building design and construction process

Optimization and Multi-benefits from Single Strategies

- Degree of design integration and synergistic outcomes

Local Building Industry Culture and Bidding Climate

- Availability of green materials and recycling facilities
- Nature of labor force and experience with unconventional materials, systems, and methods of construction

THE GREEN BUILDING LANDSCAPE AND THE ROAD AHEAD

THE TRIUMPH OF GREEN BUILDING RATING SYSTEMS AND ABOVE CODE PROGRAMS

Baseline Codes

For the purpose of this chapter, the term "baseline codes" is used to describe all building related codes, such as those produced by the International Code Council (ICC), other than the International Green Construction Code (IgCC). Those base-linecodes include the International Building Code (IBC), International Fire Code (IFC), International Mechanical Code (IECC), International Plumbing Code (IPC), International Existing Building Code (IEBC), International Energy Conservation Code (IECC), International Zoning Code (IZC), International Code Council Performance Code (ICCPC), International Residential Code (IRC), International Fuel Gas Code (IFGC), and International Wildland-Urban Interface Code (IWUIC). See Figure 3-1.

Building in accordance with standard construction practice, or baseline codes and regulations alone, has proven in the long run to be detrimental to our natural environment and long-term well-being. That is not to say that standard practices and baseline codes and standards are inappropriate or should be abandoned. In fact,

Figure 3-1

ICC family of baseline codes.

the opposite is true. Baseline codes are the foundation upon which green buildings are constructed. In order to effectively mitigate the negative effects of the built environment on the natural environment, both baseline codes and green and sustainable codes, standards or rating systems must be reconciled and complied with.

Above, Beyond, Stretch, and Reach

Green buildings are often described as being "above or beyond code" and green building codes are sometimes referred to as "stretch or reach" codes. Green building rating systems, codes, and standards move beyond the status quo and are "above code" in the sense that their provisions either exceed current minimum requirements of baseline codes, or address sustainability related concerns that are not addressed in baseline codes. The IgCC, as we will see, is an exception to this rule and a bit of a paradox in that, although it is a code, it also exceeds the requirements of the baseline codes. Green rating systems, green codes, and green standards require or encourage the implementation of "above code" concepts and green related best practices in an attempt to ensure that buildings of the future become increasingly more environmentally responsible, and that we gradually move toward the creation of a built environment that has zero negative impact on the natural environment.

In order to truly be "above" or "beyond" code, green buildings must also comply with baseline codes. Many of the requirements of baseline codes were created to resist environmental forces. Buildings must resist rain, snow, fire, seismic activity, wind, and gravity, etc. If these, and many other principles, were not addressed, buildings would be neither durable nor sustainable. Resisting catastrophic events such as earthquakes and floods, as well as everyday environmental realities such as gravity, are important fundamental aspects of sustainable building which the baseline codes have come to address in great depth. Inadvertently, however, baseline codes have only addressed half of the equation. Although they have addressed the impact of the environment on buildings, they have failed to address the negative impact that buildings have on the natural environment. That is the missing piece of the puzzle that green and sustainable building is intended to address. The goal of building green is to begin to mitigate the negative impacts of buildings on the environment, from the process of extracting materials from the earth and their manufacturing, transportation, design, and construction to occupancy, lifetime operation and maintenance and, ultimately, their decommissioning or adaptive reuse.

A multitude of tools, in the form of green and sustainable building codes, standards, and rating systems, are now available to help move the construction industry in this direction. All of these tools address the following primary environmental impact areas:

1. Sustainable site development
2. Water conservation
3. Material resource conservation
4. Energy efficiency
5. Indoor environmental quality
6. Building operations and maintenance

In order to qualify as a green building, each of these primary environmental impact areas must be addressed. A building that only addresses water conservation, for example, may not perform any better than baseline requirements from an energy perspective and is not what has commonly come to be accepted as a green building. Marketing such a building as a green building would be considered "greenwashing." Greenwashing is the practice of making misleading claims or stating half truths in the context of green building or "greening" the environment in general. More accurately, it would be reasonable to simply say that such a building incorporates green features related only to water conservation.

Defining Green

Greenwashing

The act of misleading consumers regarding the environmental benefits of a product or service or the environmental practices of a company.

Before green building became a mainstream topic, a number of jurisdictions and local builder associations developed their own green building programs to meet their local environmental concerns. In the United States, pioneering municipalities such as Austin, Texas and Scottsdale, Arizona, among others, created and adopted green building programs which were customized to meet local environmental agendas. Austin, considered by some to be the birthplace of green building, began its program in 1991. Scottsdale began its program in 1998.

Today, spurred on by the popularity and saturation of green building in all sectors of the industry and the marketplace, as well as the general increased recognition by the public of the critical importance of green and sustainable concepts, refined and comprehensive national green building rating systems, codes, and standards are taking the place of local and regional homegrown programs. In addition, these rating systems, codes, and standards are gaining footholds and are being adopted in jurisdictions where they have never before been considered.

Following is a brief description of some of the most noteworthy green and sustainable rating systems, codes, and standards that are currently in use in the United States.

LEED

The United States Green Building Council's (USGBC's) Leadership in Energy and Environmental Design (LEED) rating systems are a suite of rating systems intended to be applied on a *voluntary* basis. USGBC was incorporated in 1993, and the first version of LEED was published in 1998. The LEED suite of building rating systems includes separate rating systems for new construction, existing buildings including operations and maintenance, commercial interiors, core and shell, schools, retail including new and commercial interiors, healthcare, homes, and neighborhood development.

LEED is the most widely used green building rating system in the United States and has been credited with inspiring innovation, driving demand for more sustainable and environmentally sensitive buildings and communities, and changing the way that much of the building industry approaches design, construction and operations, both in the United States and internationally. USGBC has formed green building councils in other countries that use modified versions of its U.S. LEED rating systems.

USGBC is a nonprofit, nongovernmental organization whose membership includes a diverse group of stakeholders from both the public and private sectors. LEED rating systems are created with the intent to be applied on a voluntary basis.

They were never intended to be applied on a mandatory basis. USGBC, however, has been a strong supporter of mandatory green requirements in the baseline codes, has served as partners in the development of both the ASHRAE 189.1, and has advocated for the adoption of both the IgCC and ASHRAE 189.1.

LEED for New Construction and Major Renovations (LEED-NC) is a green building rating system intended to apply to commercial construction. LEED-NC 2009 is the version discussed herein. LEED-NC is divided into seven categories (see Table 3-1) and four green rating levels: Certified, Silver, Gold, and Platinum (see Table 3-2). There are prerequisites for five main environmental categories: Sustainable Sites, Water Efficiency, Energy and Atmosphere, Materials and Resources, and Indoor Environmental Quality. Prerequisites set minimum mandatory program requirements that must be met for a building to be eligible for a LEED rating and certification. A point based system is then used to encourage the implementation of other green and sustainable best practices.

Table 3-1
LEED Categories for New Construction (NC)

Category	Possible Points	Description
Sustainable sites	26	Addresses environmental concerns related to site selection, alternative modes of transportation, sustainable landscapes, protection of natural habitats, stormwater management, reducing heat island effect, and mitigating light pollution
Water efficiency	10	Encourages the use of strategies and technologies that reduce the amount of potable water consumed in buildings
Energy and atmosphere	35	Sets parameters to reduce the amount of energy required for building operation, encourage the use of more benign forms of energy such as renewable energy and the use of refrigerants with a low potential for causing ozone depletion and climate change
Materials and resources	14	Addresses resource efficiencies related to material selection, waste reduction, and waste management
Indoor environmental quality	15	Addresses occupant's health, safety, and comfort through improved ventilation, managing air contaminants, specifying low emitting materials, controllability of lighting and ventilations, and daylighting with outdoor views
Innovation in design	6	Recognizes integrated design process, exemplary performance measures, and innovative building features
Regional priority	4	Provides an incentive for strategies and design features that address geographically-based environmental priorities
Total	110	

Table 3-2
LEED Ratings for New Construction, Core and Shell, and Schools

Certification Level	Points
Certified	40–49
Silver	50–59
Gold	60–79
Platinum	80 and above

The Green Building Certification Institute (GBCI) is the certifying arm of the USGBC. It not only rates and certifies LEED buildings, it also develops and administers the LEED professional credential exams, including LEED Green Associate, LEED AP with specialty credentials and LEED for Homes Green Rater. LEED is not intended to be administered by jurisdictions, though USGBC has made special exceptions in certain cases.

Released in 2008, the LEED for Homes rating system is applicable to newly constructed residential green buildings that are three stories or less in height. USGBC has a network of regional third-party green and energy raters to evaluate project submittals and conduct inspections and performance testing for certification. LEED for Homes is targeting the top 25% of new homes with best practice environmental features. LEED for Homes rating system works by requiring a minimum level of performance through prerequisites, and rewarding improved performance in each of its environmental categories. Similar to LEED-NC, there are four green rating levels: Certified, Silver, Gold, and Platinum.

IgCC

The International Code Council's (ICC) International Green Construction Code (IgCC) is a green building code composed primarily of minimum mandatory requirements. It is not a rating system. However, in a new twist for codes, it also includes a limited number of jurisdictional and owner/designer choices somewhat reminiscent of rating systems. This new regulatory framework offers options that allow consideration of local environmental and geographic conditions. The IgCC is written in mandatory language and is intended to be adopted on a mandatory basis, but can also be adopted on a voluntary basis. It was produced in accordance with ICC's government consensus process in cooperation with ASTM International and the American Institute of Architects (AIA). The IgCC addresses new construction and additions, renovations and additions to existing buildings, except that it does not address low-rise residential buildings, in a single document.

IgCC Public Version 1.0 was published in March of 2010, IgGG Public Version 2.0 was published in November of 2010 and the 2012 IgCC will be published in March of 2012. It has been adopted in two states and a handful of municipalities at the time of the writing of this book. Unless otherwise noted, this book addresses IgCC Version 2.0.

The IgCC is an overlay code that interfaces with and follows the organizational framework of other I-Codes. It is intended to be administered by code officials through the established regulatory process of performing building plan review, inspections, and special inspections by third-party entities. Thus, as nation-wide system of verifiers is already in place. As code officials already must familiarize themselves with construction documents during the plan review process, conduct on-site field inspections, and accept third-party verification certificates, they may be the most cost-effective body to administer green and sustainable requirements.

The vast majority of the IgCC is comprised of minimum mandatory requirements that address green best practices. The IgCC, like other International Codes, mandates that its requirements be implemented under specific conditions, using triggers such as the occupant load and building and the use or occupancy classification to determine when the application of various criteria is reasonable and must be complied with. However, the IgCC differs from other codes in that it also requires that jurisdictions select compliance options that are appropriate for their region and owners select electives that are appropriate for their project.

Jurisdictional options are selected upon adoption by filling in Table 302.1 (see Table 3-3). The decisions selected by the jurisdiction in this table then become mandatory for all buildings constructed in the jurisdiction. Most provisions listed in Table 302.1 are there because they are not suitable as mandatory requirements in all jurisdictions. For example, jurisdictions can choose to allow building on "Greenfield" sites only where certain infrastructure, such as mass transportation, is located nearby. This is widely considered a green practice. Some jurisdictions, however, may have no mass transportation available, and adopting a code with a mandatory provision such as this might effectively prohibit all building within the

Table 3-3
IgCC Requirements Determined by the Jurisdiction

Section	Section Title or Description and Directives	Jurisdictional Requirements	
Chapter 3. Jurisdictional Requirements and Project Electives			
302.1 (2)	Optional compliance path–ASHRAE 189.1	☐ Yes	☐ No
302.1 (3)	Project Electives–The jurisdiction shall indicate a number between 1 and 14 to establish the minimum total number of project electives that must be satisfied.	_____	
Chapter 4. Site Development and Land Use			
402.2.3	Conservation area	☐ Yes	☐ No
402.2.5	Agricultural land	☐ Yes	☐ No
402.2.6	Greenfields	☐ Yes	☐ No
402.3.2	Stormwater management	☐ Yes	☐ No
403.4.1	High occupancy vehicle parking	☐ Yes	☐ No
403.4.2	Low emission, hybrid and electric vehicle parking	☐ Yes	☐ No
405.1	Light pollution control	☐ Yes	☐ No
Chapter 5. Material Resource Conservation and Efficiency			
502.1	Minimum percentage of waste material diverted from landfills.	☐ 50% ☐ 65% ☐ 75%	
Chapter 6. Energy Conservation and Earth Atmospheric Quality			
Table 602.1, 302.1, 302.1.1	*zEPI* of Jurisdictional Choice – The jurisdiction shall indicate a *zEPI* of 46 or less in Table 602.1 for each occupancy for which it intends to require enhanced energy performance.	See Table 602.1 and Section 302.1	
602.3.2.3	Total annual CO_2e emissions limits and reporting	☐ Yes	☐ No
613.2	Post Certificate of Occupancy zEPI, energy demand, and CO_2e emissions reporting	☐ Yes	☐ No
Chapter 7. Water Resource Conservation and Efficiency			
702.1.2	Enhanced plumbing fixture and fitting flow rate tier.	☐ Tier 1 ☐ Tier 2	
702.7	Municipal reclaimed water.	☐ Yes	☐ No

Table 3-3 (Continued)

Section	Section Title or Description and Directives	Jurisdictional Requirements	
Chapter 9. Commissioning, Operation and Maintenance			
904.1.1.1	Periodic reporting	☐ Yes	☐ No
Chapter 10. Existing Buildings			
1006.4	Evaluation of existing buildings	☐ Yes	☐ No
Appendices			
Appendix B	Greenhouse gas reduction in existing buildings	☐ Yes	☐ No
B103.1	Compliance level–the jurisdiction to select phases only where "Yes" is selected in the previous row.	☐ Phase 1 ☐ Phase 2 ☐ Phase 3 ☐ Phase 4	
B103.2	Where "Phase 1" is selected under Section B103.1–jurisdiction to indicate the number of months to be used in association with Section B103.2.	_____ months	
B103.3	Where "Phase 2" is selected under Section B103.1–jurisdiction to indicate the number of years and the percentage to be used in association with Section B103.3.	_____ years _____ %	
B103.4	Where "Phase 3" is selected under Section B103.1–jurisdiction to indicate the number of years to be used in association with Section B103.4.	_____ years	
B103.5	Where " Phase 4" is selected above–jurisdiction to indicate the number of years and the percentage to be used in association with Section B103.5.	_____ years _____ %	
Appendix C	Sustainability measures	☐ Yes	☐ No
Appendix D	Enforcement procedures	☐ Yes	☐ No

jurisdiction. Jurisdictions must carefully analyze the implications of their selections in Table 302.1. Alternately, if they find Table 302.1 too restrictive or difficult to comprehend, they can decide not to select any additional requirements in the table. Because even in this form, the core of the code is composed of minimum baseline requirements that are not tied to Table 302.1, the IgCC is poised to significantly reduces the negative impact of buildings on the natural environment.

Table 302.1 also allows jurisdictions to require higher performance in certain areas, such as water and energy, in accordance with their regional environmental priorities. In some regions, such as the southwest, water may be a major concern, where as in other areas, it may not be a top priority.

Jurisdictions may also select ASHRAE 189.1 as a compliance option in Table 302.1. Where the ASHRAE 189.1 option is selected, ASHRAE 189.1 replaces Chapters 4 through 11 of the IgCC. This arrangement prevents jurisdictions from being responsible for the burden of administering two complete and complex codes that address similar issues, but do so in intrinsically different ways.

In addition, Table 302.1 requires that each jurisdiction choose between 1 and 14 "project electives" as the minimum number of above IgCC baseline measures the jurisdiction choose a number between 1 and 14 as the minimum number of "project electives", which must be complied with on each project. Project electives are chosen by the owner or design professional from those listed in Table 303.1. The project electives used can vary from project to project. Project electives encourage,

Green Codes and Standards
Intent of International Green Construction Code

"Safeguard the environment, public health, safety and general welfare through the establishment of requirements to reduce the negative potential impacts and increase the positive potential impacts of the built environment on the natural environment and building occupants, by means of requirements related to: conservation of natural resources, materials and energy; the employment of renewable energy technologies, indoor and outdoor air quality; and building operations and maintenance."

Source: Section 101.3, International Green Construction Code, Public Version 2.0, November 2010

but do not require, the consideration and implementation of green and sustainable best practices that are otherwise diffi cult or impossible to mandate. For example, awarding credit for building upon "Brownfi eld" sites encourages the reclamation of contaminated sites, which is generally considered to be a green practice. However, it would be unrealistic to require that all buildings be constructed on Brownfi eld sites. Project electives, like rating systems, also encourage the construction of higher performance buildings. For example, one project elective is awarded for each 10% improvement in energy performance above baseline IgCC requirements. The result is that ten credits are earned whenever building energy is net-zero, meaning that the building produces as much energy as it consumes (see Table 3-4).

Table 3-4
IgCC Project Electives Checklist

Section	Description	Check the Corresponding Box to Indicate Each Project Elective Selected	Jurisdictional Determination of Non-availability
Chapter 3. Jurisdictional Requirements and Project Electives			
304.1	Whole Building Life Cycle Assessment (LCA)	☐ (5 Electives*)	☐
Chapter 4. Site Development and Land Use			
407.2.1	Flood hazard avoidance	☐	☐
407.2.2	Agricultural land	☐	☐
407.2.3	Wildlife corridor	☐	☐
407.2.4	Infill site	☐	☐
407.2.5	Brownfield site	☐	☐
407.2.6	Existing building reuse	☐	☐
407.2.7	Greenfield development	☐	☐

Table 3-4 (*Continued*)

Section	Description	Check the Corresponding Box to Indicate Each Project Elective Selected.	Jurisdictional Determination of Non-availability
407.2.8	Greenfield proximity to development	☐	☐
407.2.9	Greenfield proximity to diverse uses	☐	☐
407.2.10	Native plant landscaping	☐	☐
407.2.11	Site restoration	☐	☐
407.3.1	Changing and shower facilities	☐	☐
407.3.2	Long-term bicycle parking and storage	☐	☐
407.3.3	Preferred parking	☐	☐
407.4.1	Site hardscape 1	☐	☐
407.4.2	Site hardscape 2	☐	☐
407.4.3	Site hardscape 3	☐	☐
407.4.4	Roof covering	☐	☐
407.5	Light pollution	☐	☐
Chapter 5. Material Resource Conservation and Efficiency			
508.2	Waste management (502.1 + 20%)	☐	☐
508.3(1)	Reused, recycled content, recyclable, bio-based and indigenous materials (70%)	☐	☐
508.3(2)	Reused, recycled content, recyclable, bio-based, and indigenous materials (85%)	☐ (2 Electives*)	☐
508.4.1	Service life – 100 year design life category	☐	☐
508.4.1	Service life– 200 year design life category	☐ (2 Electives*)	☐
508.4.2	Interior adaptability	☐	☐
Chapter 6. Energy Conservation, Efficiency, and Earth Atmospheric Quality			
613.3	zEPI reduction project electives		
613.3	Project zEPI is at least 5 points lower than required by Table 302.1	☐	☐
613.3	Project zEPI is at least 10 points lower than required by Table 302.1	☐ (2 Electives)	☐
613.3	Project zEPI is at least 15 points lower than required by Table 302.1	☐ (3 Electives)	☐
613.3	Project zEPI is at least 20 points lower than required by Table 302.1	☐ (2 Electives)	☐
613.3	Project zEPI is at least 25 points lower than required by Table 302.1	☐ (4 Electives)	☐
613.3	Project zEPI is at least 30 points lower than required by Table 302.1	☐ (5 Electives)	☐
613.3	Project zEPI is at least 35 points lower than required by Table 302.1	☐ (6 Electives)	☐

(*Continues*)

Table 3-4 (*Continued*)

Section	Description	Check the Corresponding Box to Indicate Each Project Elective Selected.	Jurisdictional Determination of Non-availability
613.3	Project zEPI is at least 40 points lower than required by Table 302.1	☐ (8 Electives)	☐
613.3	Project zEPI is at least 45 points lower than required by Table 302.1	☐ (9 Electives)	☐
613.3	Project zEPI is at least 51 points lower than required by Table 302.1	☐ (10 Electives)	☐
613.4	Mechanical systems	☐	☐
613.5	Service water heating	☐	☐
613.6	Lighting systems	☐	☐
613.7	Passive design	☐	☐
Chapter 7. Water Resource Conservation and Efficiency			
710.2.1	Fixture flow rates are one tier above that required by Table 302.1	☐	☐
710.2.1	Fixture flow rates are two tiers above that required by Table 302.1	☐ (2 Electives*)	☐
710.3	On-site wastewater treatment	☐	☐
710.4	Non-potable outdoor water supply	☐	☐
710.5	Non-potable water for plumbing fixture flushing	☐	☐
710.6	Automatic fire sprinkler system	☐	☐
710.7	Non-potable water supply to fire pumps	☐	☐
710.8	Non-potable water for industrial process makeup water	☐	☐
710.9	Efficient hot water distribution system	☐	☐
710.10	Non-potable water for cooling tower makeup water	☐	
710.11	Graywater collection	☐	☐
Chapter 8. Indoor Environmental Quality and Comfort			
809.2.1	VOC emissions–flooring	☐	☐
809.2.2	VOC emissions–ceiling systems	☐	☐
809.2.3	VOC emissions–wall systems	☐	☐
809.2.4	Total VOC limit	☐	☐
809.3	Views to building exterior	☐	☐
809.4	Interior plant density	☐	☐

* Where multiple electives are shown in the table in the form "(x electives)", "x" indicates the number of credits to be applied for that elective to the total number of *project electives* required by the jurisdiction as shown in Section 302.1(3) of Table 302.1.

ASHRAE 189.1

In 1975, ASHRAE developed the first standard in the nation for energy efficiency requirements in buildings. In early 2010, ASHRAE released the 189.1 Standard for the Design of High-Performance Green Buildings Except Low-Rise Residential Buildings.

Green Codes and Standards
The Purpose of ASHRAE 189.1 Standard—Design of High-Performance Green Buildings

The purpose of this standard is to provide minimum requirements for the siting, design, construction, and plan for operation of high-performance green buildings to:

a. balance environmental responsibility, resource efficiency, occupant comfort and well-being, and community sensitivity, and

b. support the goal of development that meets the needs of the present without compromising the ability of future generations to meet their own needs.

ASHRAE 189.1 is a standard produced in accordance with ANSI consensus guidelines, and in cooperation with USGBC and the Illuminating Engineers Society (IES). It is intended to be adopted on a mandatory basis and is written in mandatory language, but can also be adopted on a voluntary basis. It contains minimum requirements for green and sustainable building and is not a rating system. It establishes baseline requirements for new commercial high-performance green buildings in the areas of site sustainability, water-use efficiency, energy efficiency, indoor environmental quality, material resources, and construction and operations.

Where ASHRAE Standard 90.1 sets the foundation for energy efficiency, Standard 189.1 builds on that foundation with additional energy efficiency measures and renewable energy requirements, pushing the building industry towards net-zero-energy buildings. The standard covers all nonresidential buildings and all residential spaces in buildings that are more than three stories in height. Within these buildings, Standard 189.1 applies to new and renovated buildings and their systems, and new or renovated portions of buildings and their systems. The standard includes mandatory provisions in each section with an option to use a prescriptive path or performance path for compliance. The prescriptive option provides a simple way to show compliance that involves little or no calculations. The performance option contains an alternate way to show compliance that provides more design flexibility and is typically more complex than the prescriptive option.

As mentioned previously, ASHRAE 189.1 is included as a commercial compliance option in the IgCC. Where ASHRAE 189.1 is selected in Table 302.1 of the IgCC, it replaces Chapters 4 through 11 of the IgCC. IgCC and ASHRAE 189.1 are the first comprehensive model codes and standards for green building in the United States.

ICC 700

First published in 2008, the National Association of Home Builder's (NAHB's) ICC 700 National Green Building Standard is a rating system produced in accordance with ANSI consensus guidelines. The NAHB Research Center serves as the Secretariat in the ANSI process, while ICC is a cooperating partner.

ICC 700's scope includes all green residential occupancies, from single-family to high-rise residential buildings, including the residential portions of mixed-use buildings. In addition to being applicable to the construction of new residential buildings, ICC 700 also has provisions for rating green subdivision development and green renovations and additions to existing residential buildings.

ICC 700 is both a point based rating system and a standard. Much like LEED, it contains four performance levels based on a wide array of optional green practices and has very few mandatory items or prerequisites. Unlike LEED and most other rating systems, ICC 700 requires that the minimum number of points in each environmental category (water, energy, etc.) be ramped up at each performance level. Thus buildings that achieve higher performance levels in ICC 700 tend to perform better in all environmental categories as compared to other rating systems. In other rating systems, green point based practices used to qualify for higher threshold levels can be taken from any category, meaning those practices that are the easiest or most cost effective are often chosen rather than those that are the most environmentally beneficial (see Table 3-5).

The standard is intended to be administered by any "Adopting Entity," defined in the standard as a governmental jurisdiction, green building program or any other third-party compliance body that adopts the standard and is responsible for its administration. As such an Adopting Entity, NAHB has updated its NAHB Green *Program* to be based upon the provisions of ICC 700. This program is administered by the NAHB Research Center and has certified green residential buildings across the nation on a voluntary basis using its network of local NAHB certified verifiers. NAHB Green also has a very elegant and effective online tool to give preliminary indications of compliance with the standard and program. This tool can be accessed at www.NAHBGreen.org/ScoringTool.aspx.

Table 3-5
ICC 700 Threshold Point Ratings for Green Buildings

Green Building Categories			Performance Level Points[1][2]			
			Bronze	Silver	Gold	Emerald
1.	Chapter 5	Lot Design, Preparation, and Development	39	66	93	119
2.	Chapter 6	Resource Efficiency	45	79	113	146
3.	Chapter 7	Energy Efficiency	30	60	100	120
4.	Chapter 8	Water Efficiency	14	26	41	60
5.	Chapter 9	Indoor Environmental Quality	36	65	100	140
6.	Chapter 10	Operation Maintenance and Building Owner Education	8	10	11	12
7.		Additional Points from any Category	50	100	100	100
		Total Points:	222	406	558	697

(1) In addition to the threshold number of points in each category, all mandatory provisions of each category shall be implemented.
(2) For dwelling units greater than 4,000 square feet (372 m²), the number of points in Category 7 (Additional Points from any category) shall be increased in accordance with Section 601.1. The "Total Points" shall be increased by the same number of points.

Green Code and Standards
California Green Building Standards Code

The purpose of this code is to improve public health, safety, and general welfare by enhancing the design and construction of buildings through the use of building concepts having a positive environmental impact and encouraging sustainable construction practices in the following categories:

1. Planning and design
2. Energy efficiency
3. Water efficiency and conservation
4. Material conservation and resource efficiency
5. Environmental air quality

CALGreen

First published in 2008 and becoming mandatory in 2011, the California Green Building Standards Code (CALGreen) is the first statewide green building code in the nation and establishes minimum green building standards for residential, commercial, institutional, and publicly funded building projects throughout the state of California. The 2010 edition contains both mandatory standards and voluntary measures set at higher environmental threshold levels. A local jurisdiction may enact ordinances establishing more restrictive or higher environmental measures based on local climate, geological, topographical, or environmental conditions. At the time of this writing, no other state in the United States has adopted green requirements for both government and private-sector buildings, and both residential and nonresidential buildings, on a mandatory basis. Though CALGreen may not address green and sustainable issues to the depth that some other codes, standards, and rating systems do, the fact that it is applicable to all buildings constructed in the state on a mandatory basis means that it will undoubtedly produce significant environmental benefits on a scale that, to date, no other state has addressed.

Green Globes

Green Globes is a voluntary green building rating and management system administered by the Green Building Initiative (GBI) in the United States. Green Globes originated in Canada, having been derived from the Building Research Establishment's Environmental Assessment Method (BREEAM) in 1996. BREEAM is a green and sustainable building rating system that was initiated in the United Kingdom in 1990. The "Green Globes" title was first used in 2000. GBI acquired the rights to use Green Globes in the United States in 2004. The discussion herein relates to GBI's United States version of Green Globes.

Green Globes for New Construction (NC) and Continual Improvement for Existing Buildings (CIEB) utilize a web-based application for evaluating and rating the environmental benefits of different design scenarios for new buildings and major renovations. As a questionnaire driven program, users are guided through a sequence of questions at each stage of the design process that provide guidance for integrating elements of building sustainability. The Green Globes survey tool automatically generates reports to help optimize energy savings, reduced environmental impacts, and lower maintenance costs.

The Green Globes rating system measures environmental building performance on a 1,000 point scale divided by seven categories, with each having an assigned number of points (see Table 3-6).

Building projects that have scored a minimum threshold of 35% of the 1,000 available points are eligible for a third-party review of documentation and an on-site inspection and formal Green Globes certification. Buildings can achieve a Green Globes rating of one to four Green Globes (see Table 3-7).

Green Globes is similar to LEED-NC in many respects, but addresses some additional areas including project management, emergency response planning, durability, adaptability, deconstruction, life-cycle assessment, and noise control. See

Table 3-6
Green Globes Rating System

Assessment Category	Points	Description
Project Management	6%	Design process, environmental purchasing, commissioning
Site	11.6%	Ecological impact, development area, watershed features, enhancement
Energy	38%	Performance, efficiency, demand reduction, energy efficient features, use of renewable energy, transportation
Water	10%	Performance, conservation, treatment
Resources	10%	Low impact materials (LCA), re-use, demolition, durability, recycling
Emissions and Effluents	7.6%	Air emissions (boilers), ozone depletion, water and sewer protection, pollution controls
Indoor Environment	20%	Ventilation, lighting, thermal, and acoustical comfort, ventilation system
Total Points	1000	

Table 3-7
Green Globes Ratings

Percentage of Points	Globes Level	Description
85–100%	Tier 4	Reserved for select buildings that serve as national or world leaders in reducing environmental impacts and efficiency of buildings
70–84%	Tier 3	Demonstrates leadership in energy and environmentally efficient buildings and a commitment to continual improvement
55–69%	Tier 2	Demonstrates excellent progress in reducing environmental impacts by applying best practices in energy and environmental efficiency
35–54%	Tier 1	Demonstrates movement beyond awareness and a commitment to good energy and environmental efficiency practices

Table 3-8
Green Globes Assessment Categories

1. Project Management, Policies and Practices (50 pts)

- Integrated design process
- Environmental purchasing
- Commissioning (plans for systems testing after construction)
- Emergency response plan

2. Site (115 pts)

- Development area
- Ecological impacts (erosion, heat island, light pollution)
- Watershed features
- Site ecology enhancement

3. Energy (360 pts)

- Energy performance
- Reduced demand (space optimization, microclimatic design, daylighting, envelope design, metering)
- Energy efficiency features (lighting, heating, and cooling equipment).
- Renewable energy (solar, wind, biomass, etc)
- Transportation

4. Water (100 pts)

- Water performance
- Water conserving features (equipment, meters, irrigation systems)
- On-site treatment (stormwater, graywater, blackwater)

5. Resources (100 pts)

- Low-impact systems and materials (LCA)
- Minimal use of non-renewables
- Reuse of existing buildings
- Durability, adaptability, and disassembly
- Demolition waste (reduce, reuse, recycle)
- Recycling and composting facilities

6. Emissions, Effluents, and Other Impacts (75 pts)

- Air emissions (boilers)
- Ozone depletion
- Sewer and waterway protection
- Pollution control (procedures, compliance with standards)

7. Indoor Environment (200 pts)

- Ventilation system
- Indoor pollution control
- Lighting (daylighting and electric)
- Thermal comfort
- Acoustic comfort

COURTESY OF THE GREEN BUILDING INITIATIVE (GBI)

Table 3-8. Unlike LEED-NC, if points are not available to a project, they do not count in the total achievable points. For example, LEED penalizes projects that do not build on a brownfield or that are not in close proximity to transit stops or mixed-use services. Under the LEED rating system, the available number of points is fixed,

Green Codes and Standards
The Intent of ANSI/GBI 01 Green Building Assessment Protocol for Commercial Buildings

- Use less energy
- Conserve water resources
- Emit fewer pollutants (such as greenhouse gases, airborne pollutants, liquid effluents, and/or solid waste)
- Provide a healthier indoor environment for occupants

whereas under Green Globes, the total potential number of points is adjusted, depending on the project's location. Green Globes rates the work of the project team and does not address issues that are outside of the team's control, such as the location of the project. Another important difference is that LEED has a number of prerequisites or mandatory requirements for each of the rating categories, whereas Green Globes has none.

GBI 01

The GBI 01 Green Building Assessment Protocol for Commercial Buildings was derived from the Green Globes assessment rating system for New Construction (see Table 3-8). Like Green Globes, GBI 01 is a rating system that is intended to be used on a voluntary basis for the construction of new buildings. It is not intended to be used on a mandatory basis, or for alterations, renovations or additions to existing buildings, nor is it intended to be applied to residential buildings over three stories in height. The GBI was the first green building organization to commit to taking a commercial building rating system (Green Globes™) through the America National Standards Institute (ANSI) consensus-based process. It was first published in 2010, and it is that version that is discussed herein.

Like the Green Globes rating system, GBI 01 addresses seven areas of assessment: Project Management, Site, Water, Energy, Emissions, Indoor Environment, and Resources. Unlike Green Globes, the standard contains minimum point requirements in each of the seven areas of assessment, ensuring the building attains a minimum level of sustainability in these areas. The energy section includes minimum achievement levels and carbon equivalency measures that are used in combination with energy performance goals. The standard offers the user either a performance path or a prescriptive path (Path A and B, respectively). Path A offers a higher potential credit total (300 points), and incorporates a methodology that considers total carbon dioxide equivalent emissions in its analysis and requires determination of the energy use intensity of the proposed building, which is then benchmarked through the use of the ENERGY STAR Target Finder program. Path B offers an alternative approach to earn energy performance credit, up to 250 points.

Buildings that complete the required GBI administered third-party assessment process, achieve a minimum overall percentage of the applicable points, and meet the minimum requirements in each assessment area are awarded certification. The levels of achievement for certification range from one to four, based on the percentage of applicable points scored, as verified by the third-party site assessment. For a further description of GBI 01 levels, see the discussion under Green Globes.

FEDERAL INITIATIVES

Various actions have been taken at the federal level to encourage commercial green building. The U.S. Office of Management and Budgets Circular A-11 was produced in 2002. It encouraged federal agencies to apply the LEED rating system or U.S. Department of Energy's (DOE) ENERGY STAR program to all new and renovated federal buildings. Since that time, LEED has been applied to most new and renovated federal buildings. In 2005, the Energy Policy Act (EPAct) was passed. It provided increased incentives for solar and wind power and energy conservation in private sector buildings. In 2008, the Emergency Economic Stabilization Act was passed. It included energy provisions that extended many energy tax credits. In addition, it created new tax incentives that encourage increased energy efficiency in both commercial and residential buildings. State and local public utility programs also often provide incentives to conserve energy.

In 2010, the U.S. Army established a new policy to design and build efficient military construction projects using the ASHRAE Standard 189.1 (ASHRAE 189.1) as the baseline. The policy requires that facility construction projects follow specified requirements and guidance for building siting, storm water management, energy efficiency, cool roofs, reduced indoor and outdoor water use, and energy/water metering. The policy applies to all construction and renovation of new buildings and structures in the U.S. territories, permanent overseas active Army installations, Army Reserve Centers, Army National Guard facilities, and Armed Forces Reserve Centers.

The ENERGY STAR program was established in 1992 as an energy conservation program sponsored by the U.S. Environmental Protection Agency (EPA) and the U.S. DOE. It is designed to promote widespread energy efficiency improvements and is referenced in many green building rating systems in various capacities. In addition to addressing more than 35 product categories, it addresses energy efficiency in new homes and commercial and industrial buildings, including the costs and risks involved in purchasing energy-efficient equipment. It provides a rating system to assess building energy performance and validate savings. ENERGY STAR qualified new homes are at least 15% more efficient than homes built to the 2004 International Residential Code (IRC) and include other features which make them typically 20 to 30% more efficient than standard homes.

The WaterSense program was established in 2007 by the U.S. EPA, the WaterSense program seeks to protect the nation's water supply by promoting water conservation and enhancing the market for water-efficient products. WaterSense labeled products are backed by independent third-party testing and certification, and meet EPA's specifications for water efficiency and performance. The Water-Efficient Single-Family New Homes program provides a comprehensive approach for water savings including indoor plumbing fixtures and landscape irrigation.

THE 2030 CHALLENGE

The 2030 Challenge is an industry initiative that recognizes that buildings are the major source of global demand for energy and materials that produce by-product greenhouse gases (GHG). Slowing the growth rate of GHG emissions and then reversing it is the key to addressing climate change and keeping global average temperature less 2°C above pre-industrial levels. The 2030 Challenge calls upon the global design and construction community to reduce fossil fuel energy consumption in all new buildings by 60% immediately (see Figure 3-2) as compared to the regional average for each building type, and incorporates tiered stepped targets, with the ultimate goal of achieving carbon-neutral buildings by 2030 (using no fossil fuel GHG-emitting energy to operate).

Yet only a small percentage of the world's new buildings are actually built green (only 3% were in the United States in 2007), and many "green" buildings were not any more energy efficient than required by local codes. Even LEED did not require higher energy efficiency than base codes until 2008. In light of these facts, we appear to be short of meeting the goals of the 2030 Challenge, though we are only in the first phase.

The 2030 Challenge also challenges the industry to address existing buildings. By 2035 approximately 75% of the built environment will be either new or renovated. As existing buildings make up approximately 90% of the built environment in

Figure 3-2

The 2030 Challenge to reduce fossil-fuel energy consumption.

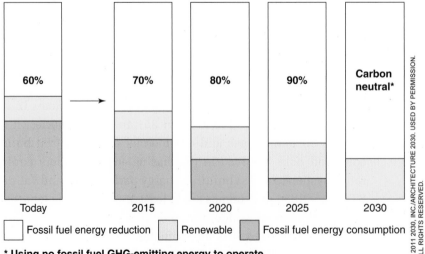

any given year, this segment of the market cannot be overlooked if 2030 Challenge goals are to be met.

PUSH, PULL, AND LEAPFROG

Green building rating systems, codes, and standards continually update their criteria with each edition in an effort to stay on the cutting edge. They push, pull, and leapfrog each other as they all move closer toward the goal of producing buildings that have zero negative impact on the natural environment. In addition, green and sustainable building requirements and best practices are migrating from one program to another (see Figure 3-3).

The Living Building Challenge is a net zero impact green building rating system. It was developed by the Cascadia, Washington State chapter of USGBC. It may come closer to producing truly sustainable buildings than any other rating system, code or standard currently available. New technologies are being implemented and tested in such buildings. The Living Building Challenge is pulling from the top. Given current technologies and realities, however, there are limitations and barriers to the goal of producing buildings that truly have zero environmental impact. Some of these technologies come at a cost premium. In fact, some of the technologies needed to reign in and tame our misuse of old technologies and the effects of them have yet to be developed. Programs like the Living Building Challenge foster such progress and serve as think tanks and test grounds for their development. Those experiences and lessons will then filter down to more widely accepted and used voluntary rating systems such as LEED and Green Globes, where they will be further tested and improved upon. As the actual performance, environmental benefits and feasibility of these new green and sustainable best practices come to be realized, they may eventually become "code ready," and be proposed for inclusion in codes and standards intended to be adopted on a mandatory basis, such as the IgCC and ASHRAE 189.1. Some of these practices may then migrate further, becoming mandatory requirements in the base codes.

Figure 3-3

The expanding scope of codes and standards push voluntary green building rating programs while simultaneously being pulled by voluntary green building rating programs resulting in a trend towards zero impact buildings.

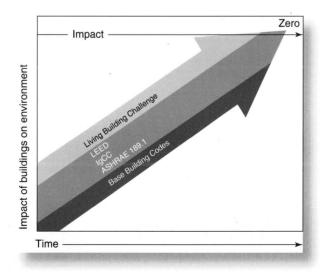

It is important to note that base codes are virtually always adopted on a mandatory basis, while green and sustainable codes and standards, at least at present, are not. Moving requirements into the base codes inherently means that they become mandated and can no longer simply be ignored, as voluntary requirements can.

Urban Green, a green and sustainable building program in New York City that was developed by their local USGBC chapter, is actively involved in moving green requirements, a handful at a time, directly into the city's mandatory base codes. These requirements will then affect all subsequent construction in the city.

Currently, programs like the Living Building Challenge are directly affecting relatively few buildings. In contrast, base codes that incorporate sustainable requirements, as well as green and sustainable codes and standards like the IgCC and ASHRAE 189.1 where they are adopted on a mandatory basis, have the ability to raise the floor of sustainability for all buildings. Thus they are capable of achieving results of a truly massive scale and magnitude that is not possible with voluntary and experimental green programs alone.

As all of these rating systems, codes, and standards continue to push and pull and leapfrog each other forward, it is the mandatory adoption and implementation of green and sustainable codes and standards such as the IgCC and ASHRAE 189.1 that will ultimately ensure that we will, eventually, virtually eliminate the negative effects of the built environment upon the natural environment.

Until we reach that point, however, voluntary rating systems will continue to play an important role, even where mandatory programs based on documents such as the IgCC and ASHRAE 189.1 are in place: voluntary rating systems will continue to encourage and acknowledge higher performance that exceeds the baseline requirements of the IgCC and ASHRAE 189.1. In the meantime, in order to both raise the floor of sustainability and pull it forward from the top, many jurisdictions will find it quite appropriate and beneficial to adopt both mandatory codes and standards which create baseline requirements, as well as voluntary rating systems which encourage higher performance and foster innovation.

Comparing Rating Systems, Codes, and Standards

In the United States, LEED green and sustainable rating systems are currently the most widely used of all green and sustainable building provisions for all buildings except low-rise residential buildings. ICC 700 is the most widely used rating system for low-rise residential applications. LEED for Homes, Green Globes, and GBI-01, while not as widely used, are also rating systems. Within each of the aforementioned primary environmental impact areas, rating systems provide an array of green strategies or best practices that are designed to reduce the negative impacts of buildings on the environment, with points assigned to each of these green strategies.

Unlike most codes and standards, which are based on minimum mandatory requirements, green building rating systems have relatively few mandatory requirements. Green Globes has none (see Figure 3-4). Rating systems also contain few detailed requirements, and instead often reference or defer to building codes and standards for such information. Most provisions of green building rating systems

Figure 3-4

Ratio of mandatory to elective or point based provisions

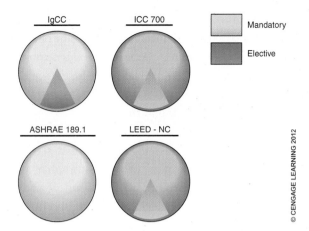

© CENGAGE LEARNING 2012

are applicable only when the owner or applicant chooses to comply with them. Whether various provisions of a rating system are implemented in a particular project become a choice for designers, contractors, and owners.

Unlike baseline codes, green building *rating systems* award points whenever one of their provisions are implemented in a building. Generally, the more provisions complied with, the higher the certification or performance level awarded within a given rating system. This system inherently encourages and rewards greener, more sustainable and higher performance buildings. Multiple thresholds also allow for "entry level" green buildings and recognize and encourage the construction of "exemplary," as LEED puts it, green buildings or buildings which achieve the highest environmental performance levels.

Most green and sustainable rating systems, however, are much better at encouraging the implementation of a greater number of green practices than they are at accurately rating the environmental impact of buildings, comparing buildings, or measuring building performance, though they are getting better. Most rating systems prior to 2008, for example, did not weight each of their best practices in relation to their ability to actually mitigate negative environmental impact. Accurate weighting is very difficult to evaluate and accomplish and, more of than not, is a much more subjective process than most would consider acceptable. That is why some rating systems offer the same amount of credit for each of the best practices that they encourage. ICC 700, GBI-01, and LEED-H, however, do weight their practices at present, as does the 2009 version of LEED-NC. While not a pure rating system, the IgCC uses project electives to encourage higher performance beyond its minimum requirements, and bundles practices in and effort to achieve some balance or weighting between best practices.

Note the use of the word "encourage." Rating systems encourage, but do not require, the consideration and implementation of many green and sustainable practices, leaving the choice as to which practices are actually implemented to the owner/applicant. Very few provisions or best practices are mandatory in most rating systems, and none are required in Green Globes. Even where green building practices can be easily and cost-effectively implemented, they are not identified as prerequisites or requirements in rating systems.

On the other hand, although green codes and standards, as well as ASHRAE 189.1, generally *require* or *mandate* minimum levels of performance wherever the cost/benefit is reasonable, they do not recognize or *encourage* higher performance. In green codes and standards, for example, if a building has a concrete foundation, it must be constructed in accordance with the codes or standards applicable requirements. In a green building rating system, however, the presence of a concrete foundation does not necessarily mean that it must be constructed in accordance with the provisions of the rating system. Under most rating systems, the owner and designer can choose to implement other sustainable options. The IgCC is a hybrid code that, although it is primarily composed of provisions that mandate minimum levels of performance, also contains a limited number of elective provisions that encourage and recognize higher performance.

In rating systems, designers, builders, and owners have the power to elect or choose which provisions they intend to comply with and are free to ignore others. Unfortunately, this can also mean that similar buildings that qualify for the same performance level in any given rating system, may have achieved that level by means of completely different sets of green practices. This can make it difficult to impossible to compare green buildings, even within the same rating system.

Although the intent of rating systems is to rate and compare the environmental impact of buildings, assumptions must sometimes be made which are not readily quantifiable. For example, what point value should be assigned to a materials effect on indoor air quality versus its effect on energy efficiency, for instance, may not be clear cut. Thus, some of the criteria used to evaluate green building areas can be subjective and may be based on perceived values in the context of a particular region, industry or intended user. Such criteria can vary significantly from one rating system to the next. Should points be awarded to recycled wood products, but subtracted if they contain formaldehyde? Is positive encouragement enough, even though it allows many practices which are not green and sustainable? Each green building rating system and program has its own unique approach to these issues, and each continues to evolve and improve. Seldom, however, do they introduce negative points for inappropriate environmental practices or characteristics. Instead, rating systems typically encourage the implementation of as many environmentally beneficial best practices as possible. Mandatory requirements, on the other hand, can totally eliminate the implementation of practices that are not environmentally responsible, or require that green practices be implemented where it is simply reasonable and cost effective to do so. For this reason, even rating systems usually have some mandatory requirements. LEED's "prerequisites," for example, are mandatory requirements.

Comparing Green Buildings and the Codes, Standards, and Rating Systems that Regulate or Rate Them

When a jurisdiction is attempting to choose a green building code, standard or rating system as the basis for its green building initiative, it is prudent to compare their provisions and the results that they are likely to achieve in each of the six primary environmental impact areas outlined earlier in this chapter. Comparisons of specific buildings as applied to each code, standard or rating system, or using one code, standard or rating system as a benchmark for all others, are usually misleading and yield skewed results. Instead, a level playing field should be established that reflects the environmental goals of the jurisdiction.

It is the environmental goals of the jurisdiction in each of the primary environmental impact areas (site, natural resources, materials, energy, water, indoor environmental quality, and atmospheric air quality), and the ability of each of the codes, standards, or rating systems to ensure that these goals are met, that should be the basis for comparison. Even where there is a lack of specific jurisdictional environmental goals, it is the ability of each of the codes, standards, or rating systems to achieve predictable, reliable, and environmentally beneficial results in each of the these primary environmental impact areas that should be assessed and compared. Ease of use, enforceability, and administration should also be considerations. In addition, care should be taken to ensure that undue hardships are not created that severely limit construction in the jurisdiction. Such hardships may simply ensure that the program is short-lived. For example, restricting building only to locations that are within very close proximity to mass transit or prohibiting construction in foodplains may totally prohibit construction in some jurisdictions. In some cases, it may be necessary to amend the code or standard upon adoption to adequately address such issues.

Realistically, comparisons of the environmental performance of green buildings are valid only between buildings that have been evaluated under the same rating system. And even within the same rating system, two buildings qualifying for the same certification level may have achieved their "points" by implementing vastly different practices. This is due to the fact that most provisions in rating systems are elective, meaning that buildings can use widely varying practices to achieve their certification or performance level. For example, after the minimum number of points are earned in each of the primary environmental impact areas in LEED for the Certified level, points required for the higher Silver, Gold, and Platinum levels can be acquired from any one of the primary environmental impact areas. This, again, makes it difficult to compare buildings within the rating system, because they can use widely varying practices to achieve their rating. Additional points for higher performance thresholds can be acquired from any environmental category in LEED. This problem is addressed to some degree in ICC 700 by requiring that performance in each of the primary environmental impact areas be ramped up to qualify for each of its Bronze, Silver, Gold, and Platinum performance levels. Green Globes provides an elegant solution, by simply scoring or rating each of the primary environmental impact areas separately.

Comparisons of buildings evaluated under different rating systems are often misleading and ambiguous at best, since each rating system assigns different values to each of its provisions (has different scoring methodology), has some differing provisions, has different minimum and mandatory requirements, and has different requirements for ramping up performance at each of their respective performance or threshold levels (i.e., Silver, 2 Globes, etc.). Quite often, a specific building may score higher under rating system "A" versus rating system "B," yet comparisons of another building might yield the opposite result. In addition, buildings that outwardly appear to be very similar, even if evaluated in accordance with the same rating system, may actually attain a vastly different rating. Again under a rating system, where a green building has a concrete foundation, that does not necessarily mean that the foundation must be built to green foundation standards. Rating systems typically have few mandatory provisions and, instead, are filled with a multitude of provisions from which the designer, builder or owner can choose to implement at their own discretion. Green building codes and standards that are composed of primarily mandatory requirements have a better chance of producing predictable and more consistent results, though they may not drive or encourage higher performance as rating systems do. That is why the IgCC incorporates aspects of mandatory base codes, as well as a limited number of elective provisions much like rating systems. The IgCC also allows jurisdictions to make choices that reflect the environmental goals of the jurisdiction. No other green building code, standard or rating system incorporates the ability to set regional environmental priorities in a jurisdiction as the IgCC does.

To further put comparisons of green buildings in perspective, note that some buildings that score low or do not even qualify for the lowest tier in many rating systems may be more environmentally responsible than those that do qualify at their highest levels. For example, almost any small building, particularly if it produces its own energy (i.e., is a net-zero energy building), even though it may not implement principles pertaining to sustainable site, water conservation or indoor air quality principles, could quite feasibly be more environmentally responsible than a large green building serving a similar function. A grass hut may have very low environmental impacts, but would not score well under most rating systems.

If it is difficult to compare buildings, even within the same rating system, as you can imagine, it is much more difficult to compare the relative advantages, disadvantages and environmental effectiveness of various codes, standards, and rating systems. The nature of the choices available and the points assigned to each provision can vary significantly between them. Comparisons of rating systems based on the number of points the percentage of points required in each primary environmental impact are can also be misleading. For example, where a rating system has ramped up performance requirements for each of its primary environmental impact areas (water, materials, energy, etc.), as the ICC 700 National Green Building Standard does, or rates each of these areas separately, as Green Globes does, the relative balance of point requirements between these areas becomes less significant, at least when buildings are compared within that rating system.

To summarize, when states and municipalities are evaluating and comparing various green and sustainable building codes, standards, and rating systems for possible use as the basis for their green building programs, comparisons of the minimum requirements of each rating systems in each of the primary environmental categories (site, sand, water, material, energy, IEQ, etc.) are usually most relevant. It is most effective to look at the general results that are likely to be produced by the application of the rating systems in each of the primary environmental impact areas, as opposed to relying on data related to the application of each rating system in the context a few specific buildings.

The minimum requirements that virtually any building must meet in each of the primary environmental impact areas of each comparison code standard or rating system by performance threshold levels should be analyzed. Local requirements to achieve higher performance levels, especially if the program is to be mandatory, should be analyzed carefully to ensure that hardships are not created. What should ultimately be compared is whether meeting the provisions of the comparison rating systems (at the minimum level required by the jurisdiction) is likely to meet the environmental goals of the jurisdiction. Occasionally, it may be necessary modify various provisions upon adoption to address some of these issues. The EPA's Sustainable Design and Green Building Toolkit for Local Governments was created to help jurisdictions move toward sustainability and evaluate some of the relevant criteria in green and sustainable codes, standards, and rating systems. It is available at www.epa.gov/region4/recycle/green-building-toolkit.pdf.

Mandatory versus Voluntary Programs and Adoptions

Transforming the market into one in which green building becomes standard practice often begins with voluntary green building programs and rating systems. These types of programs have been used to open the door and set the stage for the transition to mandatory green building code provisions. As voluntary programs become more widely accepted, the design and construction communities become educated with regard to the costs, benefits, and practical implementation of green and sustainable design, construction practices, and systems operations. This often breaks down the barriers to the adoption of green and sustainable codes, standards, and rating systems.

When a green building program is not part of a governmental initiative, it is always voluntary. When a state or municipality implements a green building program, it is usually voluntary, but it can be mandatory, or become mandatory only under specific conditions. The determination of whether a program is to be mandatory or voluntary is not a function of the rating system that is used as the basis of the program. It is a direct function of the nature of the adoption by the jurisdiction. It is the governmental jurisdiction itself that decides whether a program will be applied on a mandatory or voluntary basis, not the rating system, code or standard or those that developed it.

When states and municipalities first initiate green building programs, compliance is generally mandatory for buildings owned by the governmental jurisdiction itself, but voluntary for private sector buildings. Governmental jurisdictions generally do this in order to set an example of what they feel are responsible building practices, to save energy, water, maintenance, and other costs and to encourage the private sector to follow their lead. For the private sector, although there may be resistance to mandatory programs, there is usually little resistance to voluntary programs. This is because, of course, where there is no mandate, no one is ever actually required to comply. Usually, incentives such as expedited plan review are used to encourage participation by the private sector. Time is money to developers and builders, so this incentive is often very effective. Once design professionals, builder/contractors, and owners become familiar with green building and are aware of its advantages, there is often less resistance to making the green building program mandatory for private sector buildings. This appears to be especially true when even former skeptics, as they begin to be involved in the actual construction of green buildings, begin to realize that it is possible to comply with the program cost effectively. Although all green buildings, just as all non-green buildings, may not be always be "affordable" buildings, it is possible to construct green buildings economically. This may not always be an easy task, but the fact that even low-income housing developments have been built to the requirements of LEED demonstrates that it is feasible. In addition, many developers, contractors, builders, designers, engineers, consultants, and manufactures have successfully used green and sustainable building as a marketing tool. It can truly be a win-win scenario for all involved.

Although the number of mandatory green building programs in the United States is currently relatively small when compared to the number of voluntary programs, the trend toward mandatory programs appears to be increasing. It is interesting to note that, in India, where green building programs are implemented, they are typically mandatory, whereas compliance with building codes is typically voluntary. In the United States, representatives of the U.S. Department of Energy have indicated that the magnitude of the issues surrounding energy efficiency, a primary component of green and sustainable building, are so massive that ramped-up *mandatory* energy provisions are the only realistic means to meet energy goals that are necessary to avoid significant and potentially catastrophic environmental consequences.

Even in areas where no local green building program exists, the number of buildings being built voluntarily to green standards continues to increase each year. Green building has become a successful marketing tool for many builders and developers. Green buildings are marketed as being environmentally responsible, providing healthy environments, and implementing water and energy conservation principles that can save clients money. Certification under a green building program or rating system, or receiving a green certificate of occupancy which demonstrates compliance with the IgCC or ASHRAE 189.1, lends credibility to their claims of being green. National programs, such as LEED, Green Globes, and NAHB Green, also offer a means by which virtually any building in the country can be certified green, regardless of whether local jurisdictions have green programs in place or not.

Voluntary green building programs have and continue to pave the way and soften the resistance to mandatory programs. McGraw-Hill construction's 2011 Market Outlook predicts that green building in the United States will account for 28 to 35% of the share of new nonresidential construction in 2010, a $43 to $54 billion opportunity and more than 50% of the 2008 market share. This has also been done in a marketplace penetrated primarily by voluntary green and sustainable programs. The barriers to mandatory adoption are beginning to come down and the doors to mandatory adoption are opening.

If baseline codes were the "newcomers on the block," would they face the same obstacles and criticisms that green building rating systems, codes, and standards currently face? Possibly, someday soon, a developed world without mandatory green and sustainable codes and standards may become as unthinkable as a developed world without building, fire, and plumbing codes.

Whether green building should be voluntary or mandatory is often a topic for community and political debate, but will increasingly become a part of local economic development tools. China, for example, has embraced green building not only as an environmental necessity, but as a business opportunity.

Only mandatory adoption of green and sustainable requirements that are created and intended to be mandatory basis are equipped to revere negative trends and move towards the goal of producing a truly sustainable built environment.

HOW BASELINE CODES SUPPORT GREEN BUILDING

A building that complies only with a green building rating system or standard, but does not comply with the requirements of the baseline codes, is neither a green nor a sustainable building. There is a high degree of interdependency between green building codes, standards, and rating systems and the base codes. In order to produce truly green and sustainable buildings, both baseline codes and green codes, standards or rating systems must be complied with.

The baseline codes are, in essence, the tree on which green building ornaments are hung. If the tree on which these ornaments are hung falls, the ornaments cannot stand on their own, and all comes crashing down. Thus an effective argument can be made for virtually each and every provision in each of the I-Codes to substantiate its connection to, or support of, green and sustainable building practices.

Baseline codes are intended to ensure the safety, heath, and general welfare of the public, and to protect property and emergency responders. This information is typically found in the intent provision on the very first page of each of the International Codes. The base codes accomplish their intent by addressing structural strength, stability, sanitation, fire safety, adequate light and ventilation, and energy conservation, etc. Each of these concerns are, directly or indirectly, concerns which are also common to green and sustainable building. For example, a building that does not comply with the requirements of the base codes for resistance to fire, water intrusion, wind, and seismic activity, etc., is more likely to fail sooner, have a shorter service life, and thus is also likely to have increased negative environmental impact. Sanitation requirements support plumbing systems that, in turn, support water conservation. Energy codes are conservation codes. Light and ventilation requirements in the base codes are directly related to indoor air quality and indoor comfort. Building and fire codes ensure that buildings are safe and healthy for human occupants. Buildings constructed to base codes are more environmentally friendly and are conducive to a sustainable built environment than buildings that are not constructed to such codes.

In fact, the baseline codes already address the six primary environmental impact areas mentioned earlier in this chapter (site development, water, materials, energy, indoor environmental quality and building operations and maintenance), though only the energy and plumbing codes directly address the important conservation aspects which are critical to green and sustainable building.

Buildings that are constructed in accordance with baseline codes have provided significant benefits to society. They have greatly reduced loss of property and human life, and have also significantly increased our quality of life. None-the-less, it would be a mistake to insinuate that all code-compliant buildings are green or sustainable. That would be "greenwashing." The fact remains that the baseline codes have addressed only half of the equation. While they have accounted for how environmental forces affect buildings (i.e., resisting wind, rain, fire, seismic events, etc.), they generally have not accounted for how buildings affect the environment. Building in accordance with conventional building practices of baseline codes alone have been shown to have massive and severe unintended negative environmental impacts, with potentially catastrophic long-term environmental consequences. However, if buildings were not regulated by baseline codes, much more damage to the natural environment would have been done.

The attention to green and sustainable principles represents a major paradigm shift in building code and standards arena. Green building practices are a reaction to the realization that "accepted" construction practices of the past have, in a major way, led to our current environmental predicament, and that we must do much more if we intend to begin to build in a manner that is environmentally responsible. However, even though green codes, standards, and rating systems may encourage the implementation of many green practices, they seldom provide the detailed requirements necessary for their complete, sound, safe, and effective installation. Instead, they rely on the base codes and other design professional and manufacturer's guidelines to provide such information.

The end result is that both baseline codes and green codes, standards and rating systems must collectively account for the environment impacts of buildings (see Figure 3-5). Although baseline codes may not inherently produce green buildings, they provide fundamental support for many aspects of green and sustainable building.

Because virtually every provision in the I-Codes supports green building in some way, a comprehensive description of the green and sustainable aspects of each and every code section would require be quite a prohibitively extensive work.

Figure 3-5

The expanding scope of the codes addresses the impact of buildings on the environment.

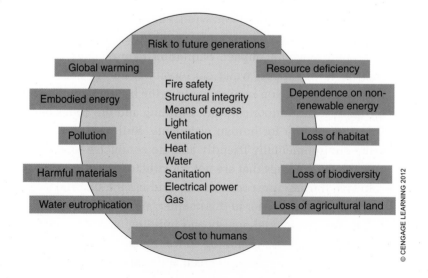

© CENGAGE LEARNING 2012

Realizing this, the information below is intended to give a brief general overview and insight into the ways in which each of the International Codes support green building. Again, this is not intended to imply that buildings constructed in accordance with the codes inherently produce green buildings, merely that the codes support green building.

Figure 3-6

International Building Code (IBC).

COURTESY OF THE INTERNATIONAL CODE COUNCIL

The International Building Code (IBC)

The IBC is applicable to all buildings and structures other than those regulated by the International Residential Code (IRC). As stated on page 1 of the IBC under "Intent," the provisions of the IBC are intended to ensure that buildings are safe and healthy for human occupancy, promote public welfare, provide protection for building contents and ensure that buildings themselves are structurally sound and stable. The IBC strives to meet these objectives through requirements that are primarily related to structural strength, durability of materials, and public safety. From a green and sustainable perspective, the IBC's requirements for structural design, exterior and interior finishes, fire-resistance rated construction and fire protection are fundamentally related to and support material resource conservation. These requirements promote building construction that is durable and resists environmental and other applicable forces, including wind, rain, snow, fire, gravity, seismic activity, and fire as well as the loads imposed by building occupants and furnishings.

As buildings which are designed and built in accordance with the IBC's requirements are designed to be capable of resisting these forces, they are much more likely to have longer useful life spans than buildings which do not comply with the requirements of the IBC. If buildings have longer effective life spans, they do not need to be replaced as often and, thereby, conserve material resources. The provisions of the IBC for structural design facilitate the design of material resource efficient structures that are also safe and structurally stable. The IBC's provisions related to durability and resistance to water intrusion, such as provisions for foundation dampproofing and exterior wall and roof coverings, moisture barriers, vapor retarders and flashing, serve to prolong the effective life span of buildings and, thereby, conserve material resources and support sustainability in the built environment. Wood is a renewable material and concrete and steel are durable materials, and the IBC and IRC provide detailed design and construction requirements for these and many other green and sustainable materials and methods. Such details are not provided in green rating systems, codes, and standards.

The IBC's requirements for accessibility ensure that buildings are useable by building occupants with a wide range of physical abilities, which is also related to a move by some green building advocates toward Universal Design (which intends to promote products and environments which are useable by all people to the greatest extent possible). Many of the IBC's requirements, such those which address means of egress (the means to leave the buildings during fire and other emergency events), fire-resistance rated construction and fire protection, are related to public safety. Buildings that are safe are likely to have longer useful life spans than those that are not.

Figure 3-7

International Fire Code (IFC).

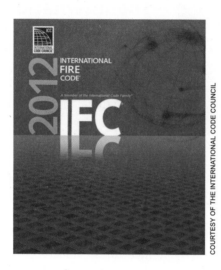

COURTESY OF THE INTERNATIONAL CODE COUNCIL

The International Fire Code (IFC)

The IFC is intended to provide life safety and property protection from fire hazards, explosions or dangerous conditions in buildings and structures, as well as to provide safety to fire fighters and emergence responders. By promoting building construction that is resistant to fire, explosions, and other dangerous conditions, including hazardous and flammable material storage and use, the IFC ensures that buildings have longer useful life spans. Buildings with longer effective life spans do not need to be replaced as often and, thereby, conserve material resources which, in turn, support sustainability in the built environment. The requirements of the IFC for fire-resistance-rated construction, interior finishes, fire protection systems, and means of egress duplicate many of those found in the IBC. See the previous discussion related to the IBC for more details related to those requirements.

Figure 3-8

International Mechanical Code (IMC).

COURTESY OF THE INTERNATIONAL CODE COUNCIL

The International Mechanical Code (IMC)

The intent of the IMC is to safeguard life, health, property, and public welfare with respect to mechanical systems which control interior environmental conditions. It strives to accomplish these goals by regulating the design, construction, installation, quality of materials, location, operation, and maintenance of mechanical systems. Primary subject areas are ventilation, exhaust systems, duct systems, combustion air, chimneys and vents, specific appliances, fireplaces, solid fuel burning equipment, boilers, water heaters, refrigeration, hydronic piping, fuel oil piping and storage, and solar systems. Chapter 23, Solar Systems, addresses solar energy collectors. The requirements of the IMC are directly related to and support the energy, indoor air quality and environmental quality and comfort provisions of green and sustainable rating systems, codes, and standards.

Figure 3-9

International Plumbing Code (IPC).

COURTESY OF THE INTERNATIONAL CODE COUNCIL

The International Plumbing Code (IPC)

The intent of the IPC is to safeguard life, health, property, and public welfare as related to plumbing equipment and systems. It strives to accomplish these goals by regulating the design, construction, installation, quality of materials, location, operation, maintenance, and use of plumbing systems. The requirements of the IPC are directly related to and support the water conservation provisions of green and sustainable rating systems, codes, and standards. Although conserving water, such as encouraged by green rating systems and standards, is not the primary purpose of the IPC, the IPC supports green and sustainable building by providing technical requirements which facilitate the safe, sound, and complete installation of plumbing systems. Green building rating systems and standards do not provide these detailed requirements. Among the IPC's more obvious green and sustainable provisions are referenced standards which no longer prohibit the use of waterless urinals and provisions for on-site gray water recycling systems. The maximum fixture flush and flow rates of the IPC are based on the 1992 EPAct which all green building rating systems and standards use as a benchmark.

Figure 3-10

International Private Sewage Disposal Code (IPSDC).

COURTESY OF THE INTERNATIONAL CODE COUNCIL

The International Private Sewage Disposal Code (IPSDC)

The IPSDC is intended to ensure public health, safety and welfare as related to the installation and maintenance of private effluent disposal systems. Many different types of sewage disposal systems are regulated by the IPSDC. Such properly designed systems reduce the negative impact of buildings on the land and its natural resources, including the protection of ground water.

Figure 3-11

International Fuel Gas Code (IFGC).

The International Fuel Gas Code (IFGC)

The intent of the IFGC is to safeguard life, health, property, and public welfare as related to fuel gas systems. It strives to accomplish these goals by regulating the design, construction, installation, quality of materials, location, operation, maintenance, and use of fuel gas systems. The requirements of the IFGC are directly related to and support the indoor and outdoor air quality provisions of green and sustainable building rating systems, codes, and standards. Many energy and service water systems rely on fuel gas as a primary energy source.

Figure 3-12

International Energy Conservation Code (IECC).

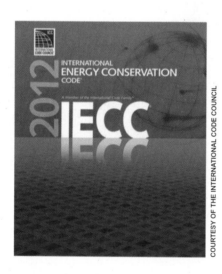

The International Energy Conservation Code (IECC)

In reaction to the 1970 oil embargo and resulting energy crisis, ASHRAE set the foundation for energy conservation with the development of Standard 90 becoming the first energy efficiency standard for buildings in the nation. The evolving standard serves as the benchmark for building codes and many commercial green building rating systems. In 1977, the first-ever energy conservation code, called the Code for Energy Conservation in New Building Construction, was developed deemed and evolved, after several adaptations, into the Model Energy Code (MEC) during the 1980s. By 1998, the model code was reintroduced by ICC as the IECC.

The intent of the IECC is to provide regulations which ensure that buildings use energy effectively. The requirements of the IECC are directly related to and support the energy conservation, fossil fuel reduction, and greenhouse gas reduction provisions of green and sustainable rating systems and standards. The IECC, as its very title states, is already a conservation code. Its baseline minimum efficiency requirements continue to be ramped up with each subsequent edition. The 2012 IECC, for example, is preliminarily estimated to produce buildings that use approximately 30% less energy than deemed buildings produced in accordance with the 2006 IECC. As such, it should be no surprise that IECC requirements have more in common with green and sustainable building provisions (albeit their energy provisions) than any other International Code. This commonality makes the IECC ripe for the direct incorporation of many green and sustainable related energy provisions. At the time of this writing, the ICC 700 committee responsible for maintenance of the document recommended that compliance with the 2012 IECC or the energy provisions of Chapter 11 of the IRC (which duplicates the residential provisions of the 2012 IECC) be deemed to comply with its minimum Bronze performance level. It has become difficult for many green rating systems, codes, and standards to stay ahead of energy codes, and they have recently begun to lag in some instances. It was not until 2008 that LEED required energy performance that exceeded the requirements of energy codes, though it did require verification of performance.

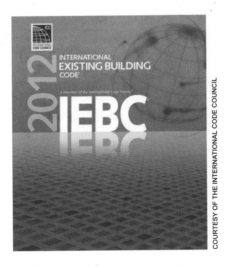

Figure 3-13

International Existing Building Code (IEBC).

COURTESY OF THE INTERNATIONAL CODE COUNCIL

The International Existing Building Code (IEBC)

The intent of the IEBC is to provide flexible alternative approaches to the International Building Code (IBC) for the repair, alteration, use, and re-use of existing buildings. As the IEBC encourages the continued use and re-use of existing buildings, it directly relates to and supports the material resources provisions of green and sustainable rating systems and standards. As the IEBC is an alternative to the IBC, much of the information in the previous discussion of the IBC is also applicable to the IEBC. In fact, in many of its requirements, especially those related to change of occupancy, additions, and where portions of existing buildings are altered, the IEBC references various IBC requirements.

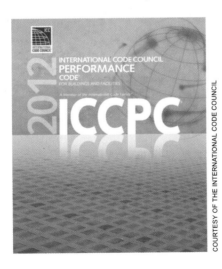

Figure 3-14

International Code Council Performance Code (ICCPC).

COURTESY OF THE INTERNATIONAL CODE COUNCIL

The International Code Council Performance Code (ICCPC)

The ICCPC is a performance-based alternative to the IBC. As such, the information contained in the previous discussion of the IBC is also applicable to the ICCPC. Unlike the IBC's "performance" based requirements, which typically specify a reference standard that must be complied with or a particular minimum value or metric that must be attained, the "performance" based requirements of the ICCPC contain few metrics and are essentially statements of the conceptual or theoretical goals that each of its provisions are intended to achieve. With chapters related to reliability and durability, stability, moisture, and interior environment, the ICCPC is pertinent to the evaluation of materials, systems, and methods of construction utilized in green and sustainable building. The ICCPC is intended to present only performance goals and not the details for compliance. The ICCPC is often used as a guide to determine what type of information should be provided building officials to facilitate the approval of innovative materials and methods or, as they are known and addressed in Chapter 1 of most International Codes, "alternative materials and methods."

Figure 3-15

International Zoning Code (IZC).

The International Zoning Code (IZC)

The intent of the IZC is to safeguard heath, property and public welfare by regulating the location, use, and occupancy of buildings and structures as related to land development. Among other functions, the IZC facilitates the establishment of a Planning Commission. The Planning Commission, in turn, is responsible for establishing a comprehensive plan which addresses land growth and use, commercial and industrial uses, transportation and utilities, community facilities, housing, environmental concerns, and geologic and natural hazards. With regards to green and sustainable building, these issues are primarily related to sustainable sites and land and natural resource management. They are also related to sustainable community planning, such as encouraged by USGBC's LEED for Neighborhood Development program.

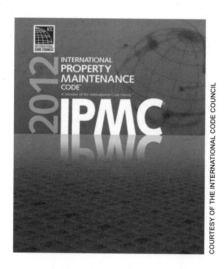

Figure 3-16

International Property Maintenance Code (IPMC).

The International Property Maintenance Code (IPMC)

Unlike the IEBC, which, as discussed previously, regulates changes in occupancy, additions and alterations to existing buildings, the IPMC serves to ensure that existing buildings and their premises are maintained and occupied in a manner which ensures public health, safety, and welfare. Although in some cases the application of the IPMC can lead to the condemnation of buildings, it more typically promotes the maintenance and continued safe operation and occupancy of existing buildings. The continued operation and maintenance of existing buildings is also addressed by green and sustainable building rating systems and standards. However, while the IPMC is primarily driven by safety and health concerns that typically arise only when a building has been neglected or improperly occupied, green and sustainable provisions related to existing buildings are primarily driven by the desire to maintain buildings and their systems in a manner which is consistent with their original design intent, such as by means of post-occupancy commissioning. Both the IPMC and the building operation and maintenance provisions of green and sustainable building rating systems and standards promote sustainability in that they facilitate the extension of the useful life span of buildings.

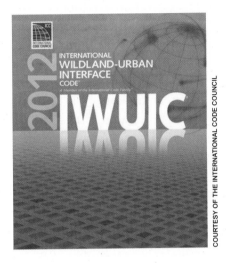

Figure 3-17

International Urban-Wildland Interface Code (IUWIC).

The International Urban-Wildland Interface Code (IUWIC)

The intent of the IUWIC is to reduce the risk to life and structures from wildland fire exposures and fire exposures from adjacent structures. Where wildland fires pose a threat, structures that comply with the IUWIC are less susceptible to damage from such exposures. The end result is that such structures are likely to have longer life spans, which relates to and supports the material resource conservation principles of green and sustainable building.

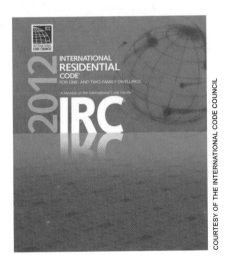

Figure 3-18

International Residential Code (IRC).

The International Residential Code (IRC)

The IRC is a "stand alone" code in the sense that all the provisions of all of the other I-Codes that are applicable to single-family dwellings, two-family dwellings and townhouses have been incorporated into this single document. The exceptions to this rule are that the IRC contains no provisions related to the IZC, IUWIC, ICCPC, and IPMC. As such, the information conveyed in the previous discussions in this book under the IECC, IFC, IMC, IFGC, and IPC are also relevant here. In addition, the first ten chapters of the IRC address many principles that are common to the IBC, so the information in the previous discussion of the IBC is also relevant to the IRC.

WHY CAN'T WE SIMPLY INCORPORATE GREEN BUILDING PROVISIONS INTO BASELINE CODES?

In some ways, incorporating green building provisions into the current building codes would be analogous to eliminating the energy code and incorporating its provisions into other baseline building codes. Since many green building provisions address buildings from perspectives that are beyond the intent and scope of baseline codes, it may make sense to group them in a separate set of provisions that interfaces with the baseline codes.

However, there are other green related provisions that may be well suited for integration in the base codes. For example, the IPC and the plumbing provisions of the IRC already address water conservation in a limited manner. Both the IPC and IRC set maximum flush and flow rates for plumbing fixtures and address the sizing of water lines. The IMC also addresses green building related issues including, but not limited to, fresh air ventilation, optimum sizing of heating/cooling systems and efficient air distribution systems. The IECC, as its very title indicates, is a conservation code. The fact that green building rating systems and standards often encourage the use of plumbing fixtures and fittings with even lower maximum rates and more energy efficient buildings, etc., will continue to pull the baseline codes in a more sustainable direction. It is possible that some green and sustainable provisions in the baseline codes may someday be pulled to their reasonable or practical limits. Even then, green and sustainable codes, standards, and rating systems may still be beneficial in that they recognize and encourage performance that is beyond code minimum in all areas related to sustainability.

The codes, however, including the IgCC, do not allow "innovative" practices which are not addressed in the code, unless evidence is provided by the permit applicant, to the satisfaction of the code official, that demonstrates that it performs in a manner that is equivalent to a material or method that is addressed in the code. Evaluation reports provided by ICC Evaluation Services are sometimes used for this purpose, as are other tests and reports. The codes do not award innovation for innovations sake. The codes protect public welfare and ensure that materials and methods perform as they are intended.

CAN GREEN BUILDINGS COMPLY WITH BASELINE CODES?

Many assume that green buildings are buildings that utilize innovative or alternative materials and methods of construction, such as rammed earth, cob, autoclave aerated concrete, wind power generation, and green roofs, etc. While this is true of some green buildings, the vast majority of green buildings simply use higher performance versions of conventional construction materials and methods. Most green and sustainable buildings are not alternative buildings in a code sense, as they do not require the application of the codes provisions for alternative materials and methods, as found in Chapter 1 of most International Codes. For example, higher efficiency building envelopes, furnaces, structural systems and plumbing systems are typically found in green buildings. These green features often simply exceed the codes minimum applicable requirements for those components and easily comply with or exceed those requirements. Thus most green features are rather easy to approve.

That said, as the built environment continues to push the envelope and move further toward producing buildings which have zero negative environmental impact, many new and innovative products and systems are likely to require evaluation under the provisions of the base codes for alternative materials and methods.

Just as the established materials and methods of construction were held to the high standards of the codes to protect the public health, safety, and welfare, so will the materials and methods of the future. It is certainly possible, as has been demonstrated many times over, that buildings can be both safe and sustainable.

THE ROLE OF CODE OFFICIALS WITH RESPECT TO GREEN BUILDING

Code officials have not typically been directly involved with the review and verification process for green buildings, except where they are responsible for the administration of a local green building program or where the evaluation and approval of alternative materials and methods of construction is required. This is because programs such as LEED, NAHB Green, and GBI 01 each have their own independent verification processes that do not involve code officials. Where buildings in a jurisdiction are voluntarily built in accordance with the provisions of an independent green rating system, since green is intended to be "beyond code," the green features of such buildings are typically well in excess of the code requirements and pose few problems for the code official. Although most green buildings do not incorporate features that must be evaluated under the codes provisions for alternative materials and methods, some do. And as the envelope of green building pushes forward towards zero negative environmental impact, the likeliness increases that green features requiring alternative approval will be encountered; at least until the codes catch up and incorporate provisions for those features.

Where there have been problems related to the approval of green features and practices, they have often stemmed from the fact that the only approval path available for a particular green feature or practice is as an alternative material or method, or that an incorrect assumption was made by owners or designers that, if a practice meets the requirement of a green rating system, it no longer must meet the requirements of the base codes. The later, of course, is not true. Green and sustainable requirements overlay the requirements of the base codes, much like the requirements of energy codes do, but they do not replace or eliminate the applicable requirements of those base codes. In fact, it is not uncommon for multiple International Codes to regulate different characteristics of a building component. For example, the R-Value of insulation may be regulated by the IECC or IgCC, but the flame spread and smoke-developed requirements for the insulation are regulated by the IBC and IFC.

Where code officials are responsible for the administration of a green building program, the extent of their green related duties are closely tied to the specific nature of the rating system or standard on which their system is based, and to whether their program is voluntary, mandatory, or some hybrid of the two. There may also be differing scenarios for residential vs. commercial buildings and public vs. private buildings.

Mandatory programs inevitably imply that a large number of projects could be affected, unless the program becomes mandatory only in specific circumstances, such as at a certain square footage threshold, for specific building types, in designated zoning districts, or only for publicly funded facilities. In general, voluntary programs are likely to garner far less participation as compared to mandatory programs, even where incentives such as expedited plan review are provided, so the workload for the building department is not likely to be as intense.

As touched upon earlier, some local green building programs, such those which reference LEED rating systems, Green Globes or GBI 01, are not intended to be administered by local jurisdictions. Instead, they must be administered by the developer of the rating system or an agency they recognize and possibly certify. In such cases, the local code official may have some extra paperwork to shuffle, but no major additional responsibilities.

The ICC 700 National Green Building Standard can be administered by any "Adopting Entity," which can be a governmental jurisdiction, third-party consultant or other compliance assurance body. The IgCC and ASHRAE 189.1 are intended to be adopted by jurisdictions and administered by building officials, or third parties authorized by the jurisdiction. As mentioned earlier, some jurisdictions have also developed and administered their own locally developed custom green provisions. In all such cases, the jurisdiction can decide to use in-house personnel to administer the program, or can contract the work out to independent consultants.

The advent of green building codes, standards, and rating systems, and their potential, if not inevitable, adoption by jurisdictions, signifies the dawn of new era for building officials. As an extension to the charge of public health and safety and general welfare, the environmental impacts of buildings is added under the scope of the IgCC. As David Eisenberg, Director of the Design Center for Appropriate Technology (DCAT) has put it, building officials are poised to become "stewards of the environment." As the IgCC evolves to become a commonly adopted code, building officials will be required to have a much wider knowledge base.

The administration of green building provisions, similar to building code requirements, requires attention in both plan review and on-site inspection phases. Training, certification and technical workbooks and commentary are required to support the adoption of green codes and standards, just as they are for the base codes. There will certainly be a learning cure for all involved in green and sustainable building. In time, however, many green and sustainable related tasks will become second nature. In an effort to shorten this learning curve, ICC currently offers seminars, workbooks, and certification related to the IgCC, with free downloadable synopsis available online for the public versions, and commentaries to follow. USGBC, ASHRAE, NAHB, ASHRAE, and others also offer similar training and support.

THE EVOLVING ROLE OF BUILDING REGULATIONS

From site development to building materials, indoor environmental quality and energy efficiency, building codes continue to play a vital role in setting an industry-wide baseline for safe, healthy, and durable buildings. Regulatory agencies across the country have a vested interest and public obligation to ensure buildings are designed and constructed to safeguard life, health, and public welfare. Since the launching of the ICC codes and the LEED rating system in 2000, there is a greater public awareness of the interdependence of the natural and built environment. Along with the growing imperative for energy independence, the scope, and intent of codes have evolved to address the environmental and health impacts associated with the design, construction, and long-term operation of buildings. Assessment tools and standards continue to be developed that evaluate the environmental impact and performance of buildings, which in turn provides a stronger green framework for evolving building codes and regulations.

Early green building program adopters have proven that alternative materials, technologies, and methods of construction can be implemented without compromising health, safety, and public welfare. ICC, ASHRAE, USGBC, ASTM, GBI, NAHB, and other building codes and standards development bodies continue this trend and have made concerted efforts to reconcile the perceived conflict between building safety and sustainability. Early green building program adopters have proven that alternative materials, technologies and methods of construction can be implemented without compromising health, safety, and public welfare.

Chapter | # SUSTAINABLE COMMUNITIES

4

SUSTAINABLE PLANNING AND SITE LOCATION

Planning is about organizing resources and making choices to achieve goals and objectives. It also involves avoiding problems before they happen. Planning for sustainable communities can help to avoid or minimize air and water pollution, help prevent loss of natural habitats, aid in the conservation of farm and forest lands, and assist in the mitigation of transportation impacts.

The location of a building site will impact greenhouse gas (GHG) emissions and other pollutants based on available options for alternative modes of transportation (see Figure 4-1), including pedestrian proximity to basic services and residential neighborhoods. The intent of responsible planning and land use is to support smart growth planning, low-impact development, protect environmentally sensitive lands (see Figure 4-2), minimize site light pollution, and manage on-site stormwater through reuse and infiltration.

Figure 4-1
Light rail train.

COURTESY OF ANTHONY FLOYD

Open spaces of nature in cities provide many of the same benefits as those just listed. Public spaces such as parks often serve as an environmental oasis in the heart of a concrete jungle. Central Park in New York City is a refuge from the commotion and environmental void of the city (see Figure 4-3). Parks and nature preserves, to a great extent, serve as places for quiet contemplation and an appreciation of nature.

Land-use and zoning ordinances are driven by local growth, economic priorities, and political concerns. Local planning agencies usually address building uses, setbacks and heights, density, transportation, and development patterns. Sustainable development practices vary greatly across the country. One city may be a leader in stormwater management, and another an innovator in traffic impact mitigation but neither may be advanced in ecosystem preservation and heat-island mitigation. Regardless of local planning incentives, every project should consider the environmental context of the site and region.

COURTESY OF ANTHONY FLOYD

Figure 4-2

Nature preserve.

Figure 4-3

Central Park, New York City.

COURTESY OF ANTHONY FLOYD

ECOSYSTEM SERVICES

As discussed in Chapter 2, an ecosystem is the set of organisms living in an area, and their interaction with each other and their physical environment.[1] Goods and services flowing from natural ecosystems are greatly undervalued by society. For the most part, the benefits of ecosystems such as clean air are not traded in the marketplace and do not necessarily respond to the economic conditions of supply and demand.

Ecosystem services are the conditions and processes through which natural ecosystems, and the species that make them up, sustain human life. They maintain biodiversity and the production of ecosystem goods, such as food, timber, biomass fuels, natural fiber, and many pharmaceuticals, industrial products, and so forth. Ecosystem services are the foundation for life-support functions, such as cleansing air and mitigating negative impacts and regeneration of valuable resources[2] (see Table 4-1).

Ecosystem services are the foundation of daily life. In a healthy natural ecosystem, soils, vegetation, and water function together in processes that provide all the goods and services

Defining Green

Ecology

Study of the relationship of all living things, including people, to their biological and physical environments. An understanding of ecology reveals how we interact with each other and our natural and built environments.

Table 4-1
Ecosystem Services

- Sun (harmful UV ray) protection
- Climate stabilization
- Temperature and other "acts of Mother Nature" moderation
- Flood and drought reduction
- Air and water purification
- Soil generation and renewal
- Crops and natural plant pollination
- Agricultural pest control
- Seed and nutrient distribution
- Biodiversity maintenance, on which much medicine, agriculture, and industry is based
- Support of human cultures
- Esthetic beauty and intellectual stimulation

© CENGAGE LEARNING 2012

---

[1]Daily, Gretchen, *Nature's Services* (Washington, DC: Island Press, 1997).

[2]Ibid.

on which humans depend.[3] Plants could not perform their functions of producing food or regulating our breathable air if they were not supported in their growth by healthy soil and clean water. Water in turn, would not be cleansed without the filtering provided by healthy plants and soils. No one element in the natural world can function without the healthy functioning of the others. Healthy ecosystems are the source of the many less tangible benefits that humans derive from nature. No one type of ecosystem has a monopoly on the goods and services it can provide.[4]

Ecosystem services are generated by a complex of natural cycles, driven by solar energy, that constitute the workings of the biosphere, the thin layer near the earth's surface that contains all known organic life. The cycles operate on very different scales. Biogeochemical cycles, such as the movement of the element carbon through the living and physical environment, are truly global and reach the top of the atmosphere to deep within the soils and ocean-bottom sediments. Life cycles of bacteria, in contrast, may be completed in an area much smaller than the point on the end of a pen. The cycles all operate at very different rates. The biogeochemical cycling of carbon, for instance, occurs at orders of magnitude faster than that of phosphorus, just as the life cycles of microorganisms may be orders of magnitude faster than those of trees.

Through the hydrologic cycle, an enormous cycling of water through the earth's atmosphere, oceans, land, and biosphere shapes the world's climate, supports plant growth, and makes life as we know it possible.[5] On a well-vegetated site with healthy, porous soils, rainwater is absorbed and transpired by vegetation, or it saturates into the soil. In developed areas with paved surfaces, rainwater runs off into storm sewers and is lost to the natural hydrologic cycle. Vegetation stabilizes soil and slopes by making them resistant to erosion.[6] In turn, healthy soils allow rainwater to infiltrate, reducing not only runoff and erosion but also sedimentation and flooding. Soils also cleanse, cool, and store rainwater, and recharge groundwater.[7]

All living things on earth are built of carbon compounds. As part of the carbon cycle, carbon dioxide (CO_2) plays a large role in earth's energy balance in the atmosphere, acting like a blanket to trap radiation that would otherwise radiate away into space.[8] Through photosynthesis and respiration, plants and animals play a key role in the circulation of carbon over a time span ranging from days to thousands of years. Over the past one hundred years, the amount of CO_2 in the

[3]The Sustainable Sites Initiatives, *The Case for Sustainable Landscapes* (ASLA: 2009) - www.sustainablesites.org/report

[4, 5]The Sustainable Sites Initiatives, *The Case for Sustainable Landscapes* (ASLA: 2009) - www.sustainablesites.org/report

[6]RPC Morgan and RJ Rickson, *Slope Stabilization and Erosion Control: A Bioengineering Approach* (London: Chapman & Hall, 1995)

[7, 8]The Sustainable Sites Initiatives, *The Case for Sustainable Landscapes* (ASLA: 2009) - www.sustainablesites.org/report

atmosphere has been steadily increasing, due largely in part to the burning of fossil fuels and also to changes in land use.

Nitrogen makes up 78% of the earth's atmosphere and ranks fourth behind oxygen, carbon, and hydrogen as the most common chemical element in living tissue. Human activities have greatly increased the amount of "fixed," or reactive, nitrogen entering the nitrogen cycle—doubling it over the course of the past 100 years.[9] The increase in fixed nitrogen to the global nitrogen cycle has grave consequences for the environment, including increased atmospheric concentrations of the GHG nitrous oxide and of the nitrogen precursors of smog and acid rain. This also results in the acidification of soils, streams and lakes.

All of these cycles are ancient, the product of billions of years of evolution. They are pervasive yet unnoticed by most people in their daily lives. Human beings depend on these natural cycles for their very existence. If the life cycles of predators that naturally control most potential pests were interrupted, it is unlikely that pesticides could effectively take their place. If the carbon cycle were disrupted, rapid climate change could threaten the existence of life as we know it.[10]

LOCATING BUILDINGS

A greenfield site is one that has not been previously developed and includes undisturbed ecosystems and ecologically productive land such as forests and native habitats. As greenfield land diminishes with sprawling development, so do natural habitats and the essential natural services benefits of our regional ecosystems. Building on previously developed sites is a key element of sustainable land use. Infill development on previously developed sites has direct access to existing infrastructure, including utilities and transportation. Infill development supports transit and pedestrian connectivity while preserving greenfield sites with open space and biodiversity.

When infill development is not possible, then development should occur in a way that is sensitive to the environmental characteristics of the site. This is particularly important for development on or near floodplains, wetlands, bodies of water, native plant and wildlife habitat areas, and steep slopes. Development on floodplains and steep slopes can lead to increased erosion, destruction of habitat, and increased water pollution. Environmentally sensitive development respects the site hydrology, topography, and native habitats by minimizing the building development footprint. The disturbed areas of the site can be restored with native plants along with the removal of invasive species.

[9]The Sustainable Sites Initiatives, *The Case for Sustainable Landscapes* (ASLA: 2009) - www.sustainablesites.org/report

[10]Ehrlich, P., and H. Mooney, "Extinction, Substitution, and Ecosystem Services," *BioScience* (1983) 33: 248–254.

COMPACT DEVELOPMENT, DENSITY, CONNECTIVITY, AND MIXED USE

Compact or clustered development is essential for walkability, reduced vehicular transportation demand, and a diverse mix of building uses and services. The degree of compactness is usually measured in terms of density or intensity. Density refers to the number of residential dwelling units per acre. Intensity is measured by the floor area ratio (FAR) and refers to commercial or mixed-use buildings. It is the ratio of the total amount of building area to the total site area.

The compactness of buildings is an indicator of how far people will walk between destinations. When densities are low, driving is the preferred and most effective means of transportation because dwellings are far from schools, employment, and community services. Where densities are higher, dwellings are closer together and within walking distance of services. Higher density uses less land for the same number of people and uses infrastructure (water, sewer, electricity) more efficiently. It also supports public transit by making it more economically viable.

Neighborhoods and communities with high connectivity support the ability to walk to work, stores, school, and community amenities. Community connectivity reduces the number of automobile trips, supports biking networks, and enhances social connections. It also supports physical exercise and improves health.

Locating development near a mix of residential and commercial uses decreases the need to travel long distances for goods and services. It also reduces the environmental impacts of a long commute. A mix of uses encourages walking and cycling to meet daily needs. It saves time and energy (fuel) and avoids pollution (vehicle emissions). In a single building, the uses can be mixed horizontally or vertically, such as a ground-level store with a dwelling unit(s) on the upper level. Urban areas that have a higher population density usually encompass a diverse mix of uses.

TRANSPORTATION

The extensive use of single-occupancy vehicles and their reliance on fossil fuel negatively impacts our environment in the context of climate change, air pollution, and resource depletion. Parking facilities and roadway infrastructure for automobiles contribute to urban heat-island effect and stormwater runoff due to impervious and heat-absorbing materials, such as asphalt and concrete. The use of public transportation helps to reduce energy demand for such transportation and associated GHG emissions, as well as the space needed for parking lots (see Figure 4-4). Minimizing the amount of parking needed in a project, especially in surface parking lots, can make more land available for open space or other uses at higher development densities.

Choosing a project site location with access to public transportation reduces carbon emissions by making it feasible to design a compact, pedestrian-accessible project with minimal parking area. Another alternative to automobiles is bicycling.

Figure 4-4

Public transportation.

© ISTOCKPHOTO/GEORGE CLERK

COURTESY OF ANTHONY FLOYD

Figure 4-5

Dedicated bicycle path.

HYBRID PARKING ONLY

COURTESY OF ANTHONY FLOYD

Figure 4-6

Hybrid-only parking.

It produces no emissions, has zero demand for fossil-based fuels, relieves traffic congestion, reduces noise pollution, and requires far less infrastructure. Bicycle path networks, parking/storage areas, and shower/changing facilities encourage residents and employees to ride bicycles to work, school, basic services, and recreation (see Figure 4-5).

Low-emission, hybrid, and electric vehicles offer the possibility of reducing smog and air pollutants as well as the environmental impacts associated with oil extraction and petroleum refining. Providing preferred parking for low-emission, hybrid, and electric vehicles serves as an incentive for employees and visitors to commercial and public service facilities (see Figure 4-6).

PLANNING TOOLS AND GUIDELINES

Smart Growth

Smart growth is a planning strategy to help communities develop and grow in a way that supports environmental conservation, economic development, and community vitality. Smart growth is transit and pedestrian oriented, and favors clustered development with a mix of housing, commercial, and retail uses. It preserves open space and protects environmentally sensitive lands as public amenities. Smart growth recognizes connections between development and the quality of life. It leverages new growth to improve the community. Smart growth is closely related to New Urbanism (discussed later) and is based on the 10 principles discussed next.[11]

1. Create Range of Housing Opportunities and Choices

Providing quality housing for people of all income levels is an integral component in any smart growth strategy. Housing is a critical part of the way communities grow, as it constitutes a significant share of new construction and development. More important, however, is that it is also a key factor in determining households' access to transportation, commuting patterns, access to services and education, and consumption of energy and other natural resources. By using smart growth approaches to create a wider range of housing choices, communities can mitigate the environmental costs of auto-dependent development, use their infrastructure resources more efficiently, ensure a better jobs-housing balance, and generate a strong foundation of support for neighborhood transit stops, commercial centers, and other services.

2. Create Walkable Neighborhoods

Walkable communities are desirable places to live, work, learn, worship, and play, and therefore a key component of smart growth (see Figure 4-7). Their desirability comes from two factors. First, walkable communities are located within an easy and safe walk to both goods (such as housing, offices, and retail) and services (such as transportation, schools, libraries) that a community resident or employee needs on a regular basis. Second, by making pedestrian activity possible, walkable communities expand on transportation options, and create a streetscape that better serves a range of users—pedestrians, bicyclists, transit riders, and those driving automobiles. To foster walkability, communities must mix land uses and build compactly, and ensure safe and inviting pedestrian corridors.

3. Encourage Community and Stakeholder Collaboration

Growth can create great places to live, work, and play—if it responds to a community's own sense of how and where it wants to grow. Communities have different needs and will emphasize some smart growth principles over others: those with robust economic growth may need to improve housing choices; others that have suffered from disinvestment may emphasize infill development; newer

[11]Smart Growth Network - www.smartgrowth.org/why.php

COURTESY OF ANTHONY FLOYD

Figure 4-7

An example of a walkable community also serving bicyclists and cars.

communities with separated uses may be looking for the sense of place provided by mixed-use town centers; and still others with poor air quality may seek relief by offering transportation choices. The common thread among all, however, is that the needs of every community and the programs to address them are best defined by the people who live and work there.

4. Foster Distinctive, Attractive Communities with a Strong Sense of Place

Smart growth encourages communities to craft a vision and set standards for development and construction that respond to community values of architectural beauty and distinctiveness, as well as expanded choices in housing and transportation. It seeks to create interesting, unique communities that reflect the values and cultures of the people who reside there, and foster the types of physical environments that support a more cohesive community fabric. Smart growth promotes development that uses natural and man-made boundaries and landmarks to create a sense of defined neighborhoods, towns, and regions. It encourages the construction and preservation of buildings that prove to be assets to a community over time, not only because of the services provided within, but because of the unique contribution they make on the outside to the look and feel of a city.

5. Make Development Decisions Predictable, Fair, and Cost-Effective

For a community to be successful in implementing smart growth, it must be embraced by the private sector. Only private capital markets can supply the large amounts of money needed to meet the growing demand for smart growth developments. If investors, bankers, developers, builders, and others do not earn a profit, few smart growth projects will be built. Fortunately, government can help make smart growth profitable to private investors and developers. Because the development industry is highly regulated, the value of property and the desirability of a place are largely affected by government investment in infrastructure and government regulation. Governments that make the right infrastructure and regulatory decisions will create fair, predictable, and cost-effective smart growth.

6. Mixed Land Uses

Smart growth supports the integration of mixed land uses into communities as a critical component of achieving better places to live. By putting uses in close proximity to one another, alternatives to driving, such as walking or biking, once again become viable. Mixed land uses also provide a more diverse and sizable population and commercial base for supporting viable public transit. Mixed land use can enhance the vitality and perceived security of an area by increasing the number of people on the street. It helps streets, public spaces, and pedestrian-oriented retail again become places where people meet, attracting pedestrians back onto the street and helping to revitalize community life.

7. Preserve Open Space, Farmland, Natural Beauty, and Critical Environmental Areas

Smart growth uses the term "open space" broadly to mean natural areas both in and surrounding localities that provide important community space, habitat for plants and animals, recreational opportunities, farmland and ranch land (working lands), places of natural beauty, and critical environmental areas (e.g., wetlands). Open space preservation supports smart growth goals by bolstering local economies, preserving critical environmental areas, improving our community's quality of life, and guiding new growth into existing communities.

8. Provide a Variety of Transportation Choices

Providing people with more choices in housing, shopping, communities, and transportation is a key aim of smart growth. Communities are increasingly seeking these choices—particularly a wider range of transportation options—in an effort to improve beleaguered transportation systems. Traffic congestion is worsening across the country. Whereas in 1982 65% of travel occurred in uncongested conditions, by 1997 only 36% of peak travel occurred in such conditions. In fact, according to the Texas Transportation Institute, congestion over the last several years has worsened in nearly every major metropolitan area in the United States.

9. Strengthen and Direct Development Toward Existing Communities

Smart growth directs development toward existing communities already served by infrastructure, seeking to utilize the resources that existing neighborhoods offer and conserve open space and irreplaceable natural resources on the urban fringe. Development in existing neighborhoods also represents an approach to growth that can be more cost-effective, and improves the quality of life for its residents. By encouraging development in existing communities, communities benefit from a stronger tax base; closer proximity of a range of jobs and services; increased efficiency of already developed land and infrastructure; reduced development pressure in edge areas, thereby preserving more open space; and, in some cases, strengthening rural communities.

10. Take Advantage of Compact Building Design

Smart growth provides a means for communities to incorporate more compact building design as an alternative to conventional, land-consumptive development. Compact building design suggests that communities be designed in a way that permits more open space to be preserve, and that buildings can be constructed to allow for more efficient use of land and resources. By encouraging buildings to grow vertically rather than horizontally, and by incorporating structured rather than surface parking, for example, communities can reduce the footprint of new construction, and preserve more green space. Not only is this approach more efficient by requiring less land for construction, it also provides and protects more open, undeveloped land than would exist otherwise to absorb and filter rainwater, reduce flooding and stormwater drainage needs, and lower the amount of pollution washing into our streams, rivers, and lakes.

New Urbanism

New Urbanism is an urban planning movement designed to capture the pedestrian-based and mixed-use characteristics of pre-automobile-dominated neighborhoods. New Urbanism is about the design of communities for people, rather than automobiles. The Charter of the New Urbanism, created by Congress for New Urbanism, identifies development at three scales: the region, the neighborhood, and the building.

A region consists of cities, towns, and communities. In major urban centers, it usually is a metropolitan area such as Chicago, Phoenix, or the Los Angeles greater metropolitan area. A region shares the same climate, ecosystem, and geographic resources. When planning for a new development, one must understand its place within the larger region.

The primary building blocks of towns and cities are neighborhoods, districts, and corridors. Neighborhoods are primarily residential areas developed around common focal points such as parks, schools, or shopping areas. Districts are usually single-use areas that serve as the primary job and retail centers. Corridors are linear, mixed-use areas that are located on major transportation corridors.

The three scales of New Urbanism development are based on the principles discussed next.[12]

1. Walkability
- Most things within a 10-minute walk of home and work
- Pedestrian-friendly street design (buildings close to street; porches, windows, and doors; tree-lined streets; on-street parking; hidden parking lots; garages in rear lane; narrow, slow-speed streets)
- Pedestrian streets free of cars in special cases

2. Connectivity
- Interconnected street grid network to disperse traffic and ease walking
- A hierarchy of narrow streets, boulevards, and alleys
- High-quality pedestrian network and public realm makes walking pleasurable

3. Mixed Use and Diversity
- A mix of shops, offices, apartments, and homes on site; mixed use within neighborhoods, within blocks, and within buildings
- Diversity of people—of ages, income levels, cultures, and races

[12]New Urbanism - www.newurbanism.org

4. Mixed Housing

- A range of types, sizes, and prices in closer proximity

5. Quality Architecture and Urban Design

- Emphasis on beauty, aesthetics, human comfort, and creating a sense of place; special placement of civic uses and sites within community; human-scale architecture and beautiful surroundings nourish the human spirit

6. Traditional Neighborhood Structure

- Discernable center and edge
- Public space at center
- Importance of quality public realm; public open space designed as civic art
- Contains a range of uses and densities within 10-minute walk
- Transect planning: highest densities at town center; progressively less dense toward the edge. The transect is an analytical system that conceptualizes mutually reinforcing elements, creating a series of specific natural habitats and/or urban lifestyle settings. The transect integrates environmental methodology for habitat assessment with zoning methodology for community design. The professional boundary between the natural and man-made disappears, enabling environmentalists to assess the design of the human habitat and the urbanists to support the viability of nature. This urban-to-rural transect hierarchy has appropriate building and street types for each area along the continuum.

7. Increased Density

- More buildings, residences, shops, and services closer together for ease of walking, to enable a more efficient use of services and resources, and to create a more convenient, enjoyable place to live.
- New Urbanism design principles are applied at the full range of densities from small towns, to large cities.

8. Green Transportation

- A network of high-quality trains connecting cities, towns, and neighborhoods together
- Pedestrian-friendly design that encourages a greater use of bicycles, rollerblades, scooters, and walking as daily transportation

9. Sustainability

- Minimal environmental impact of development and its operations
- Eco-friendly technologies, respect for ecology, and value of natural systems
- Energy efficiency
- Less use of finite fuels
- More local production
- More walking, less driving

10. Quality of Life

- Taken together, these add up to a high quality of life well worth living, and create places that enrich, uplift, and inspire the human spirit.

LEED for Neighborhood Development (LEED-ND)

The USGBC LEED-ND program was released in 2010 and was jointly developed with the Congress for the New Urbanism and the Natural Resources Defense Council. LEED-ND is the most comprehensive of all the LEED green rating programs. It places emphasis on the site selection, design, and construction elements that bring buildings and infrastructure together into a neighborhood and relate the neighborhood to its landscape as well as its local and regional context. LEED-ND is not designed to compete with or replace local zoning ordinances or comprehensive planning instruments. Instead, it's a voluntary market-based tool designed to analyze whether existing local zoning and development regulations, including landscape requirements, building codes, and infrastructure plans, are compatible with sustainable development principles. LEED-ND has three environmental categories: Smart Location and Linkage, Neighborhood Pattern and Design, and Green Infrastructure and Buildings (see Table 4-2).

Table 4-2
LEED for Neighborhood Development

Smart Location and Linkage	
Focuses on site selection that minimizes the adverse environmental effects of development across several categories, including transportation, air quality and preservation of environmentally-sensitive lands or ecosystems. Urban sprawl and associated low density, segregated housing and commercial uses are discouraged. Preference is given to locations close to existing town and city centers, sites with good transit access, infill sites, previously developed sites and sites adjacent to existing developments. Selection of sites that are within or adjacent to existing development can minimize habitat fragmentation and also help to preserve areas for recreation. Remediation and reclamation of contaminated brownfield sites make them safer and can contribute to social and economic revitalization of depressed neighborhoods.	
Prerequisites	**Credits**
• Smart location	Preferred Locations
• Imperiled species and ecological communities conservation	• Brownfield redevelopment
• Wetland and water body conservation	• Locations with reduced automobile dependence
• Agricultural land conservation	• Bicycle network and storage
• Floodplain avoidance	• Housing and jobs proximity
	• Steep slope protection
	• Site design for habitat or wetland and water body conservation
	• Restoration of habitat or wetlands and water bodies
	• Long-term conservation management of habitat or wetlands and water bodies

(Continued)

Table 4-2 (*Continued*)

Neighborhood Pattern and Design

Emphasizes the creation of compact, walkable, mixed-use neighborhoods with convenient pedestrian connections to nearby communities. Compact communities provide opportunities to reduce driving and resultant emissions, conserve economic resources and help reduce the spread of low density development across a region's landscape. Public spaces, such as parks and plazas, can encourage social interaction and active recreation while helping control stormwater runoff and reducing heat island effects. Community gardens promote social interaction and physical activity while increasing access to fresh, locally grown produce. Communities with diverse housing types permit residents to live closer to their workplaces and allow families to remain in a given neighborhood as their circumstances changes.

Prerequisites	Credits
• Walkable streets	Walkable streets
• Compact development	• Compact development
• Connected and open community	• Mixed-use neighborhood centers
	• Mixed-income diverse communities
	• Reduced parking footprint
	• Street network
	• Transit facilities
	• Transportation demand management
	• Access to civic and public spaces
	• Access to recreation facilities
	• Visitability and universal design
	• Community outreach and involvement
	• Local food production
	• Tree-lined and shaded streets
	• Neighborhood schools

Green Infrastructure and Buildings

Focuses on measures that can reduce the environmental consequences of the construction and operation of buildings and infrastructure. Including certified green buildings in projects is one way to reduce negative environmental effects. Sustainable building practices reduce waste and use energy, water and materials more efficiently than conventional building practices. Site ecology damage can be minimized during construction by confining construction activities to limited areas and restricting the development footprint.

Prerequisites	Credits
• Certified green building	Certified green buildings
• Minimum building energy efficiency	• Building energy efficiency
• Minimum building water efficiency	• Building water efficiency
• Construction activity pollution prevention	• Water-efficient landscaping
	• Existing building reuse
	• Historic resource preservation and adaptive use

Table 4-2 *(Continued)*

Prerequisites	Credits
	• Minimized site disturbance in design and construction
	• Stormwater management
	• Heat island reduction
	• Solar orientation
	• On-site renewable energy sources
	• District heating and cooling
	• Infrastructure energy efficiency
	• Wastewater management
	• Recycled content in infrastructure
	• Solid waste management infrastructure
	• Light pollution reduction

CODES AND STANDARDS

Planning Regulations

There is a strong historical link between community planning and public health. City planning and zoning regulations were developed in response to the public health crisis of the late nineteenth and early twentieth centuries. As a result of rapid industrialization and overcrowding of urban centers, planning regulations required sanitary sewers to prevent cholera epidemics and zoned areas of the city to buffer residential neighborhoods from polluting industries. Planning regulations also underlie people's daily decisions, such as where people live, work, go to school, and travel for entertainment and shopping; what to eat and where; and when to socialize and be physically active.

There is increasing documentation that a strictly automobile-oriented built environment, with separated uses accessible only by car, contributes to most of the leading chronic public health problems in the United States. Limited physical activity is a primary risk factor for heart disease, stroke, diabetes, and Alzheimer's disease. Higher-density, walkable communities, transportation options, and easy access to recreation all increase physical activity, which can have positive health effects.[13]

Emissions from transportation sources are strongly linked with respiratory diseases, and automobile accidents kill over 40,000 Americans each year. Land-use decisions impact people's access to nutritious foods, health care, green public spaces and exposure to air pollution and toxic releases. Poor mental health is associated with a number of factors related to planning, including long commute times, exposure to crime, and lack of access to services and public spaces.

---

[13]Keeler, Marian, and Bill Burke, *Integrated Design for Sustainable Building* (Hoboken: Wiley, 2009).

There are a number of tools that local planning agencies use to regulate building use, location, heights, density, and pattern of development. Each jurisdiction is different but each usually utilizes a comprehensive city plan, zoning ordinances, and community master plans. Some jurisdictions use these planning tools to promote sustainable development, whereas others continue the pattern of conventional development that reinforces environmental degradation and poor community health. With increased awareness, more communities are reforming their regulations to align with the principles of sustainable development.

International Green Construction Code (IgCC)

As mentioned in Chapter 3, the IgCC provides a comprehensive set of requirements intended to reduce the negative impact of buildings on the natural environment. It is a document that can be readily used by manufacturers, design professionals, and contractors—but what sets it apart in the world of green building is that it was created with the intent to be administered by code officials and adopted by governmental units at any level as a tool to drive green building beyond the market segment that has been transformed by voluntary rating systems such as LEED and Green Globes.

The IgCC is an overlay code that relies on the foundation provided by other I-Codes to provide communities with buildings that are safe and sustainable. Rather than the past approach of creating buildings that are capable of resisting environmental forces, consideration is given to the impacts on the natural environment from forces imposed by the built environment. The IgCC, much like the IECC, is a code that regulates buildings primarily from a public welfare perspective. The IgCC is uniquely formatted not only to require the implementation of environmentally related best practices, but to encourage practices that are difficult to mandate, as well as to offer flexibility for jurisdictions to adapt to local condition and environmental priorities.

Chapter 4 of the IgCC contains requirements for both site development and land use. The planning land-use provisions are intended to preserve natural resources, mitigate transportation impacts, and minimize light pollution (see Table 4-3).

ASHRAE 189.1 Standard for the Design of High-Performance Green Buildings

The site sustainability section of ASHRAE 189.1 supports smart growth planning and protects environmentally sensitive lands. ASHRAE 189.1 includes a set of mandatory requirements that must be met for all projects, and an option for either a prescriptive or performance set of requirements to demonstrate compliance.

With respect to sustainable communities, ASHRAE 189.1 has a mandatory provision on site selection. The intent is to minimize development on greenfields and undeveloped sites. Development is required to occur on sites that have existing infrastructure, including building reuse and modifications to existing building structures. This includes development on existing greyfields or mitigated brownfield sites.

Table 4-3
International Green Construction Code (Public Version 2)

IgCC Chapter 4: Site Development and Land Use

Section 402: Preservation of Natural Resources

402.2.1 Floodplains. Buildings and building site improvements shall not be located within a floodplain.

402.2.2 Surface water protection. Buildings and building site improvements shall not be located within a buffer around a water body or within a buffer around a water body, as defined as the ordinary high-water mark of seas, lakes, rivers, streams and tributaries which support or could support fish, recreation or industrial use.

402.2.3 Conservation area. Site disturbance or development of land within 50 feet of any designated conservation areas shall not be permitted.

402.2.4 Park land. Site disturbance of development of land located within a public park shall not be permitted.

402.2.5 Agricultural land. Buildings and associated site improvements shall not be located on land zoned for agricultural purposes.

402.2.6 Greenfield sites. Site disturbance or development shall not be permitted on greenfield sites unless located within specified proximity of existing residential land (min. 8 units per acre), walking distance of basic services, transit services or existing development with pedestrian connectivity.

402.3.1 Predesign site inventory and assessment. An inventory and assessment of the natural resources and baseline conditions of the building site shall be submitted.

402.3.2 Stormwater management. Stormwater management systems, including, but not limited to, infiltration, evapo-transpiration; rainwater harvesting and runoff reuse, shall be provided and maintained on the building site. Stormwater management systems shall address the increase in runoff that would occur resulting from development on the building site.

Section 403: Transportation Impact

403.1 Walkways and bicycle paths. Not less than one independent, paved walkway or bicycle path suitable for bicycles, strollers, pedestrians, and other forms of non-motorized locomotion connecting a street or other path to a building entrance shall be provided.

403.2 Changing and shower facilities. Buildings with a total building floor area greater than 10,000 square feet and that are required to be provided with long term bicycle parking and storage in accordance with Section 403.3 shall be provided with on-site changing room and shower facilities.

403.3 Bicycle parking and storage. Long-term and short-term bicycle parking shall be provided in accordance with IGCC Table 403.3 (see Table 4-4).

403.4 High occupancy, low emission, hybrid, and electric vehicle parking. Where employee parking is provided for a building that has a total building floor area greater than 10,000 square feet and a building occupant load greater than 100, at least 5 percent but not less than 2 of the employee parking spaces provided shall be designated as preferred parking for high occupancy or low emission, hybrid and electric vehicles.

High occupancy vehicles are defined in the IGCC as vehicles which are occupied by two or more people, when arriving and departing the site where parked, for not less than 75 percent of the vehicle trips; or as otherwise defined by state or local regulation.

Low emission, hybrid and electric vehicles are defined as vehicles that achieve EPA Tier 2, California LEVII, or a minimum of EPA LEV standards, whether by means of hybrid, alternative fuel, or electric power.

(Continued)

Table 4-3 *(Continued)*

Section 405: Site Lighting
405.1 Uplight, light trespass, and glare of all exterior lighting shall be limited on the building site in accordance with the appropriate lighting zone of IGCC Table 405.1.1 (see Table 4-5) and specified uplight, backlight and glare ratings.

Table 4-4

IgCC Table 403.3: BICYCLE PARKING

Occupancy	Specific Use	Short Term Spaces	Long Term Spaces[b, c]
R-1	Hotel, motel, boarding houses	None	1 per 50 employees; not less than 2 spaces
R-2, R-3, R-4	All	None	None
A-1	Movie theaters	1 per 50 seats; not less than 4 spaces	
	Concert hall, theaters other than for movies	1 per 500 seats	
A-2	Restaurants	1 per 50 seats; not less than 2 spaces	1 per 50 employees; not less than 2 spaces
A-3	Places of worship	1 per 500 seats	
	Assembly spaces other than places of worship	1 per 25,000 square feet; not less than 2 spaces	
A-4 – A-5	All	1 per 500 seats	
B	All	1 per 50,000 square feet; not less than 2 spaces	1 per 25,000 square feet; not less than 2 spaces
F, H, S	All, except parking facilities	None	1 per 50 employees; not less than 2 spaces
M	All	1 per 25,000 square feet; not less than 2 spaces	1 per 50,000 square feet; not less than 2 spaces
S	Transit park and ride lots	None	1 per 20 vehicle parking spaces.
	Commercial parking facility	1 per 20 vehicle parking spaces	None
I-2	All	1 per 25,000 square feet; not less than 2 spaces	1 per 50 employees; not less than 2 spaces
I-1	All	None	
E, I-4	Day care	None	
E	Schools	None	1 per 10 students
Other	Outdoor recreation, parks	1 per 20 vehicle parking spaces; not less than 2 spaces	None

Table 4-5
IgCC Table 405.1.1: EXTERIOR LIGHTING ZONES

Lighting Zone	Description
1	Developed areas of national parks, state parks, forest land and rural areas
2	Areas predominantly consisting of residential zoning, neighborhood business districts, light industrial with limited nighttime use and residential mixed use areas
3	All other areas
4	High-activity commercial districts in major metropolitan areas as designated by the local jurisdiction

COURTESY OF THE INTERNATIONAL CODE COUNCIL

Development cannot occur on a greenfield site unless conditions exist that support pedestrian connectivity in the immediate area surrounding the site. These conditions include proximity to residential density (10 units per acre), 10 basic services (with 0.5 mile; see Figure 4-8), and train service (within 0.5 mile) or other adequate transit service (within 0.25 mile). Finally, development can occur on a greenfield site that is classified as agricultural, forest, or designated park land when the specific function of the building is related to the respective use of the land. In addition to site selection, Standard 189.1 limits development in flood hazard areas, fish and wildlife habitat conservation areas, and wetlands.

REGIONAL ENVIRONMENTAL PRIORITIES

Local and regional planning and land-use issues are based on evolving community values that usually involve density, intensity, open space, building heights, uses, road infrastructure, parking, utilities and accommodation for public transportation, and light pollution. Planning and land-use issues are the most difficult to standardize from community to community. There is no "one-size-fits-all" solution. Successful communities do tend to have one thing in common—a vision of the future that reflects their collective community values.

Most local jurisdictions adopt a Comprehensive or General Plan to guide growth and development policies on character and design, land use, open spaces, the natural environment, business and economics, community services, neighborhood vitality, transportation, and growth issues. This General Plan focuses on shaping the physical form of the city and is designed to be a broad, flexible document that changes as the community needs, conditions, and direction change. It is usually revised through city-initiated amendments, citizen/property-owner requests, or through referenda (citizen petition and vote). Ultimately, the decision to amend a community's General Plan is in the hands of the local elected officials.

Local policies for the conservation of energy and natural resources are of national and global importance. Local decisions about land use, environmental preservation, and transportation can have regional and global impacts, such as climate changes, air quality, or resource depletion. Planning for sustainability often requires broad state and federal policies much like the Federal Aid Road Act (1916) and Federal Housing Administration Act (1934) that incentivized state road construction and sprawling residential development. The federal government can play a pivotal role in establishing sustainable development programs

Figure 4-8

Planning for basic services within a walkable distance.

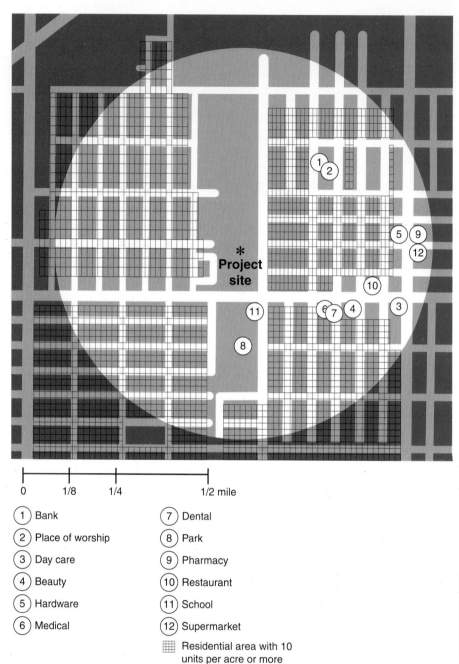

0 — 1/8 — 1/4 — 1/2 mile

1. Bank
2. Place of worship
3. Day care
4. Beauty
5. Hardware
6. Medical
7. Dental
8. Park
9. Pharmacy
10. Restaurant
11. School
12. Supermarket

Residential area with 10 units per acre or more

to protect public health and safety, directing existing state funding to encourage smart growth, and providing tax incentives for more environmentally responsible practices. Sustainable communities will be those that can manage both current and long-range planning in balance with economic growth, environmental protection, and community vitality.

SITE DEVELOPMENT

SUSTAINABILITY AND SITE DEVELOPMENT

Site development can negatively affect topography, hydrology, and local ecosystems. Site hardscape increases stormwater runoff, erosion, groundwater pollutants, and summer heat sinks. The intent of responsible site development is to protect habitats, reduce heat-island effect, minimize site light pollution, maximize pervious surfaces, retain native vegetation, and manage on-site stormwater through reuse and infiltration.

Whether a residential development, shopping center, or office building, all of these have the potential to support clean air and water, reduce flooding, cool urban air temperature, and combat climate change. Conventional land development practices too often damage the environment's ability to provide essential benefits that support human health and well-being.

Traditional landscape practices can result in nearly 90% of stormwater being lost as runoff, leading to flooding. Other impacts include the following:[1]

- Soils compacted during construction cause long-term costs in damaged vegetation and increased runoff, which leads to flooding problems and water pollution.
- Pesticides can contaminate ground and surface water, and yet more are currently applied to suburban lawns and gardens than to agricultural plots.
- Nationally, 13% of municipal waste is from yard and landscape trimmings. Such waste clogs our landfills and costs cities money.
- Landscape irrigation accounts for more than 7 billion gallons of potable water daily nationwide, at least half of which may be wasted.
- Exotic invasive species contribute to $38 billion in damage annually in the United States.

Sustainable stormwater practices (Figure 5-1) can not only reduce runoff but can reduce the use of potable water for irrigation. When we create sustainable sites, the landscape gives back:[2]

- Cleaner water and air due to ground infiltration and native vegetation
- Cooler cities as a result of reduced hardscape and heat-island effect
- Carbon capture through vegetation that mitigates climate change

---

[1,2]The Sustainable Sites Initiatives, *Landscapes Give Back* (ASLA: 2009), www.sustainablesites.org/
report.

Figure 5-1

Sustainable stormwater management practices dramatically reduce runoff.

COURTESY OF ANTHONY FLOYD

- Resource conservation and regenerative environment due to native-vegetated landscape
- Greater energy efficiency due to the shading of buildings by trees and other vegetation
- Habitat conservation and biodiversity
- Lower costs and improved performance from stormwater management
- Nature-friendly environment that produces pleasant living conditions

Sustainable development practices vary regionally based on climate, geographic region, environmental resources, and unique constraints. One region may be challenged with large-scale soil erosion and groundwater pollution while another may be faced with severe water shortages. Each project must consider the interconnected nature of environmental issues in the larger context of the region and within the framework of local planning and environmental regulatory authorities.

ECOLOGICAL PLANNING

The disconnect between buildings and nature in the industrial age was clearly articulated by Ian McHarg in his 1969 book *Design with Nature*.[3] McHarg called for environmental planning on a local level and advocated taking everything in the environment into account when planning the built environment, including rocks, soils, plants, animals, and ecosystems. It's not only about preserving—it is also about developing and managing the built environment with respect to the ecological attributes of the natural environment. Ecological planning views nature as a process and places as products of physical and biological evolution. Ecological inventory, analysis, and synthesis is vital towards understanding this process in its assessment of topography, climate, geology, soils, hydrology, vegetation, wildlife land use, and potential uses deemed suitable for the location (see Figure 5-2).

---

[3]McHarg, Ian, *Design with Nature* (Garden City: Doubleday/Natural History Press, 1969).

SITE PLANNING

Site planning must address some of the most pressing ecological challenges: a compromised ecology, an aging infrastructure, and the pressure to accommodate a growing population. Site design provides an opportunity to reduce the environmental consequences of development. Site plans should preserve the existing tree canopy and native vegetation to the extent possible while accommodating compact development. Preserving existing vegetation can reduce stormwater runoff, mitigate the urban heat-island effect, reduce the energy needed for heating and cooling, and reduce landscaping installation and maintenance costs. Trees and vegetation also reduce air pollution, provide wildlife habitat, and make outdoor areas more pleasant for walking and recreation (see Figure 5-3).

Figure 5-2
Ecological planning factors.

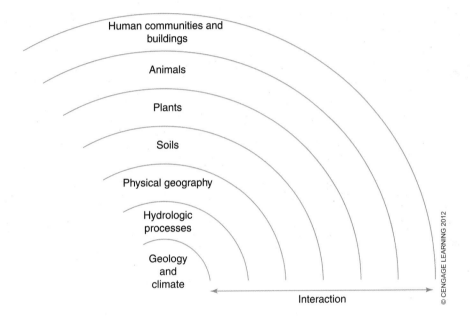

Human communities and buildings

Animals

Plants

Soils

Physical geography

Hydrologic processes

Geology and climate

Interaction

© CENGAGE LEARNING 2012

Figure 5-3

Parks and plazas serve many ecological and community purposes.

COURTESY OF ANTHONY FLOYD

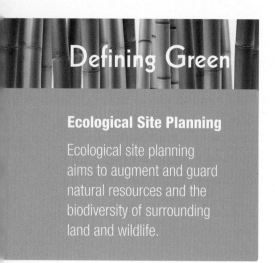

Ecological Site Planning

Ecological site planning aims to augment and guard natural resources and the biodiversity of surrounding land and wildlife.

The site design process begins as early as possible as part of the site selection and conceptual planning phase in conjunction with the following development process:[4]

1. Site analysis and programming: Property selection, stakeholder identification and outreach, information gathering, environmental inventory, conceptual planning, and development programming.
2. Preliminary planning: Initial planning of land uses, transportation networks, and major facilities; public outreach and refinement of plans.
3. Final design: Continued public outreach; preparation of final site plan, infrastructure design, and building designs; acquisition of building permits.
4. Construction and occupancy: Acceptance of site infrastructure and building by local jurisdiction and issuance of occupancy certificates by building department.

Site Inventory and Assessment

A site inventory identifies the existing conditions of the site, including topography, hydrology, soils, and ecological features. Native vegetation and site features play a vital role in supporting the local ecology. Its presence helps to prevent erosion, provides food and shelter for wildlife, and provides shade, thus reducing the urban heat-island effect. Native vegetation requires less water and maintenance than nonindigenous plant materials. In most cases, salvaging existing plant material is more economical and achieves a natural appearance in a shorter amount of time. Many trees are slow growing and can take decades to reach maturity. Factors such as the size, form, or location of certain mature plants make finding a comparable nursery-grown tree for replacement difficult to impossible. Leaving native plants in place or salvaging them for incorporation into landscaping is beneficial both from a financial and practical point of view. Protected plants should, in the most optimal situation, remain in place. Those plants that must be moved should be salvaged.

Vegetation, Soil, and Erosion Control

Removing existing vegetation disturbs soils, so controlled removal should be practiced. Without vegetation, a site loses its natural capacity for stormwater management, filtration, and groundwater recharge. Reduced vegetative cover also affects soil health, because vegetation maintains soil structure, contributes to soil organic matter, and prevents erosion.

STORMWATER MANAGEMENT

Undeveloped land has some given capacity to absorb rainfall in the soils, vegetation, and trees. The process of land development impacts the natural environment by affecting the quantity and quality of stormwater runoff on a site. Clearing of vegetation and construction of impervious surfaces reduce the capacity of the

[4]USGBC, *LEED Reference Guide for Neighborhood Development* (Washington DC: 2009).

land to absorb rainfall and increase the amount of stormwater runoff. Covering undeveloped earth with buildings and paved surfaces replaces largely pervious surfaces with impervious materials. Post-development stormwater runoff contains sediments and other contaminants that have a negative impact on water quality. Stormwater pollution sources include atmospheric deposition, vehicle fluid leaks, and mechanical equipment wastes. During storm events, these pollutants are washed away and discharged to downstream bodies of water.

Green site development seeks to protect ecosystems by reducing stormwater runoff and providing for on-site infiltration that helps maintain the natural aquifer recharge cycle. Strategies include vegetated roofs, pervious pavement, open-grid pavers (see Figure 5-4), rain gardens, vegetated swales, and infiltration basins (see Figure 5-5). Stormwater may also be collected and used (rainwater harvesting) for non-potable water purposes such as landscape irrigation, flushing toilets and urinals, and building equipment operations for evaporative cooling systems (see Figure 5-6).

HEAT-ISLAND EFFECT

Temperatures in cities are substantially higher than in surrounding rural areas due to heat-absorbing paving and building materials. Late summer afternoon temperatures in cities are 2°F–10°F (1.5°C–6°C) higher than in surrounding areas (see Figure 5-7). Only about 1% of that increase is from heat generated directly

Figure 5-4

Open-grid pavers.

COURTESY OF ANTHONY FLOYD

Figure 5-5

Infiltration basin.

COURTESY OF ANTHONY FLOYD

Figure 5-6

Storage for rainwater harvesting.

COURTESY OF ANTHONY FLOYD

by vehicles and equipment, according to Lawrence Berkeley National Laboratory (LBNL) research. The rest is from solar-heated materials and surfaces. The lack of vegetation removes the natural shading and cooling effect of trees and ground-cover materials. Heat islands contribute to higher summer cooling loads for buildings, which increases the energy consumption and the environmental impacts

Figure 5-7

Urban heat island profile.

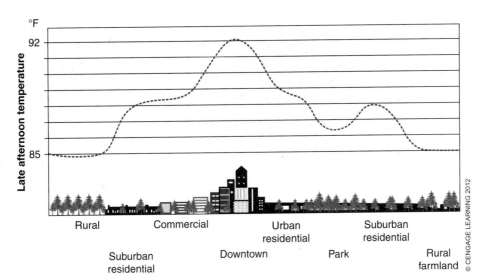

associated with producing more energy. Higher temperatures also decrease summer pedestrian activity and exacerbate health problems, because ozone (a component of smog) is created when pollutants, such as nitrogen oxides and volatile organic compounds, heat up. A temperature drop of 3°F–4°F (2°C) can reduce smog by 10%–20%, according to the LBNL.

Urban heat-island effect can be mitigated by reducing solar impact on (1) site hardscape, (2) roof surfaces, and (3) wall materials. Site hardscape includes parking lots, bike paths, walkways, courtyards, and plazas. Heat-island reduction strategies for site hardscape must minimize impervious surfaces or otherwise protect sun-exposed and heat-absorbing hardscape materials. As a stormwater management strategy, pervious paving materials have interconnected air voids and therefore have a lower capacity to absorb heat. Pervious materials include specially designed pervious concrete and asphalt, open-grid pavement, and open-graded aggregate materials (see Figure 5-8).

As an alternative to pervious paving, shading is an effective measure to reduce the heat absorption of hardscape. Shade may be provided using landscape features such as native trees, large shrubs, and noninvasive vines. Vegetation can reduce heat islands by shading buildings and pavements from solar radiation and cooling the air through evapotranspiration. Trellises and other exterior structures can support vegetation to shade parking lots, walkways, and plazas. For some climates, deciduous trees may be used to allow solar heat gain during the winter months when outside temperatures are colder, and shade during the summer months when shading is required. Shade can also be provided by shade structures including carport structures, walkway covers, plaza canopies, and adjacent buildings. Based on national standards such as ASHRAE 189.1, shading is only effective when at least 50% of the site hardscape is shaded on the longest day for the year (summer solstice) between the hours of 10 am and 3 pm.

The final approach to reduce the heat-island effect of site hardscape is to address reflectivity and emissivity. It's no surprise that light-colored, reflective surfaces heat up less in the sun. Reflectivity measures how well a material bounces

Figure 5-8

Stabilized open-graded aggregate parking area.

Defining Green

Solar Reflectance Index (SRI)

Measure of a material's ability to reject solar heat.

back radiation. But because all surfaces absorb some heat, there is also a need to consider emissivity, or how good a surface is at radiating heat back out into space. The "solar reflectance index" (SRI), as defined by ASTM E 1980, incorporates both reflectivity and emissivity. The combination of reflectivity and emissivity means that light-colored, low-emitting surfaces tend to perform better than darker, high-emitting surfaces. Effective paving materials should have a minimum SRI of 29. Many paving products have tested SRI values, including gray concrete, white concrete, and asphalt paving materials (see Table 5-1).

Another way to reduce hardscape surfaces is to locate parking under structures or the building itself. This can include using multistory or underground parking structures. Early in the design process, the project team should involve the landscape architect, architect, and civil engineer in efforts to minimize hardscape surfaces on-site, and evaluate the potential for locating parking underground or under cover.

Similar to site hardscape, roof surfaces can either absorb heat or reflect and emit heat. In accordance with most green building standards, at least 75% of a roof surface should have a minimum initial SRI of 78 for a low-sloped roof or 29 for a steep-sloped roof (see Table 5-2). Many roofing products meet the recommended minimum SRI requirements. These products may include white thermoplastic or PVC membranes, white metal roofing, and selected roofing shingles. It is important to confirm a product's SRI with the product manufacturer, as SRI may vary among product types.

Table 5-1
Solar Reflectance Index (SRI) Values for Standard Paving Materials

Material	Emissivity	Reflectance	SRI
Typical new gray concrete	0.9	0.35	35
Typical weathered* gray concrete	0.9	0.20	19
Typical new white concrete	0.9	0.7	86
Typical weathered* white concrete	0.9	0.4	45
New asphalt	0.9	0.05	0
Weathered asphalt	0.9	0.10	6

* Reflectance of surfaces can be maintained with cleaning. Typical pressure washing of cementious materials can restore reflectance close to original value. Weathered values are based on no cleaning.

Table 5-2
Recommended SRI values for Roofs

Roof type	Slope	Solar Reflectance Index (SRI)
Low-sloped roof	≤ 2:12	78
Steep sloped roof	>2:12	29

In lieu of a roof with an acceptable SRI value, a vegetated roof may be used that not only reduces heat-island impacts but also reduces stormwater impact by absorbing rainfall. A vegetated roof also improves air quality, absorbs ambient sound, serves as a thermal insulator, and provides evapotranspiration for passive cooling (see Figure 5-9). However, it is important to note that a vegetated roof may increase structural loading and waterproofing requirements. To determine whether or not a vegetated roof is appropriate for a given site, the project team should consider geographic location, annual precipitation, seasonal temperatures and sun exposures, and irrigation demand.

Vertical wall surfaces also contribute to heat retention and should be included in the reduction of the heat-island effect. The shading should be from grade level up to 20 ft (6 m) above grade on exterior walls. Similar to paving materials, east- and west-facing walls receive direct solar radiation from the low-angle sun in the morning and afternoon, which affects radiant temperatures at the pedestrian scale. This is more critical in the warmer climate zones 1 through 4. As a mitigation measure, at least 30% of the east and west above-grade walls (within a 45-degree building rotation) should be protected by either a minimum SRI value of 29 or shading with vegetation, architectural shading devices, or adjacent buildings. The difference between the SRI value of a vertical surface and that of a horizontal one is primarily in the surface film resistances. At the time of this writing, all SRI values for materials were tested only for horizontal surfaces, but these values have been considered acceptable for vertical wall surfaces by green building standards such as ASHRAE 189.1. If shading is provided in lieu of the SRI value, the shade coverage for east walls should be calculated at 10 am and on the west walls at 3 pm on the longest day of the year (summer solstice).

Figure 5-9

Vegetated roof.

COURTESY OF ANTHONY FLOYD

TRANSPORTATION IMPACT

The extensive use of single-occupancy vehicles and their reliance on fossil fuel negatively impacts our environment in the context of climate change, air pollution, and resource depletion. Parking facilities and roadway infrastructure for automobiles contribute to urban heat-island effect and stormwater runoff due to impervious and heat-absorbing materials, such as asphalt and concrete.[5] The use of public transportation helps to reduce energy demand for transportation and associated greenhouse-gas emissions, as well as the space needed for parking lots (see Figure 5-10). Minimizing the amount of parking needed in a project, especially in surface parking lots, can make more land available for open space or other uses at greater development densities.

Choosing a project site location with access to public transportation indirectly reduces carbon emissions by making it feasible to design a compact, pedestrian-accessible project with minimal parking area. Another alternative to automobiles is bicycling. It produces no emissions, has zero demand for fossil-based fuels, relieves traffic congestion, reduces noise pollution, and requires far less infrastructure. Bicycle path networks, parking/storage areas, and shower/changing facilities encourage residents and employees to ride bicycles to work, school, basic services, and recreation while providing health benefit (see Figure 5-11).

Low-emission, hybrid, and electric vehicles offer the possibility of reducing smog and air pollutants from vehicular travel as well as the environmental impacts associated with oil extraction and petroleum refining. Providing preferred parking

---

[5]USGBC, *LEED Reference Guide for Green Building Design and Construction* (Washington, DC: 2009).

Figure 5-10

Increased public transportation reduces parking area needs and opens land for other purposes.

© ISTOCKPHOTO/RYAN LANE

Figure 5-11

Proper bicycle parking/storage encourages bicycling as transportation.

COURTESY OF ANTHONY FLOYD

for low-emission, hybrid, and electric vehicles serves as an incentive for employ-ees and visitors to commercial and public service facilities (see Figure 5-12).

PLANNING TOOLS AND GUIDELINES

Sustainable Sites Initiatives

The Sustainable Sites Initiative is an interdisciplinary effort by the American Society of Landscape Architects, the Lady Bird Johnson Wildflower Center at the University of Texas at Austin, and the United States Botanic Garden. The initia-tive developed criteria for sustainable land practices that will enable built land-scapes to support natural ecological functions by protecting existing ecosystems and regenerating ecological capacity where it has been lost. The program focuses on measuring and rewarding a project that protects, restores, and regenerates eco-system services—benefits provided by natural ecosystems such as cleaning air and water, climate regulation, and human health benefits.

The new rating system gives credits for the sustainable use of water, the conser-vation of soils, wise choices of vegetation and materials, and design that supports human health and well-being. The Initiative recognizes that any landscape is capable of providing the natural benefits essential to human and ecological health. Strategically planting vegetation outdoors reduces the energy consumption needed to cool the indoors by up to 25%. In Minneapolis, street trees resulted in

Figure 5-12

Preferred hybrid parking acts as an incentive for drivers of low-emission vehicles.

COURTESY OF ANTHONY FLOYD

savings of $6.8 million in energy costs and $9.1 million in stormwater treatment, and property values increased by $7.1 million.[6] Compost and mulch can decrease soil compaction and increase soil's nutrient content and its ability to hold water.

LEED for New Construction

As mentioned in Chapter 3, LEED-NC provides a set of performance standards for certifying the design and construction phases of commercial, institutional, and high-rise residential buildings. The intent of LEED-NC is to assist in the creation of high-performance, healthy, durable, affordable, and environmentally sound buildings. The LEED-NC's Sustainable Sites section addresses environmental impacts related to the preservation of natural resources, development density, community connectivity, reducing emissions associated with transportation, heat-island mitigation, and light pollution. There is one prerequisite and 14 credits (see Table 5-3).

Green Globes for New Construction

The Green Globes Site category assessment consists of online survey questions organized under the following five stages of the design construction process: (1) Pre-design, (2) Schematic Design, (3) Design Development, (4) Construction Documents, and (5) Contracting/Construction. Each stage assesses site priorities with the objective of reducing environmental impacts (see Table 5-4).

LEED for Homes

LEED for Homes is an initiative designed to promote the transformation of the mainstream homebuilding industry toward more sustainable practices. The LEED for Home's Sustainable Sites section addresses erosion control, landscaping, heat-island effect, stormwater management, nontoxic pest control, and compact development. There are two prerequisites and 13 optional credits (see Table 5-5).

CODES AND STANDARDS

International Green Construction Code (IgCC)

As mentioned in the previous chapter, the IgCC is an overlay code that relies on the foundation provided by other International Codes to provide communities with buildings that are safe and sustainable. Rather than the past approach of creating buildings that are capable of resisting environmental forces, consideration is given

---

[6]Minneapolis Tree Advisory Commission, Minneapolis Tree Advisory Commission Annual Report, Minneapolis Park and Recreation Board, City of Minneapolis (June, 2006).

Table 5-3
LEED for New Construction: Sustainable Sites

Credit	Prerequisites and Requirements
SSp1	Construction Activity Pollution Prevention: Reduce pollution from construction activities by controlling soil erosion, waterway sedimentation, and airborne dust generation.
SSc1	Site Selection: Avoid development of inappropriate sites and reduce the environmental impact from the location of a building on a site.
SSc2	Development Density and Community Connectivity: Channel development to urban areas with existing infrastructure, protect greenfields, and preserve habitat and natural resources.
SSc3	Brownfield Redevelopment: Rehabilitate damaged sites where development is complicated by environmental contamination, reducing pressure on developed land.
SSc4.1	Alternative Transportation, Public Transportation Access: Reduce pollution and land development impacts from automobile use.
SSc4.2	Alternative Transportation, Bicycle Storage and Changing Rooms: Reduce pollution and land development impacts from automobile use.
SSc4.3	Alternative Transportation, Low-Emitting and Fuel-Efficient Vehicles: Reduce pollution and land development impacts from automobile use.
SSc4.4	Alternative Transportation, Parking Capacity: Reduce pollution and land development impacts from automobile use.
SSc5.1	Site Development, Protect, or Restore Habitat: Conserve existing natural areas and restore damaged areas to provide habitat and promote biodiversity.
SSc5.2	Site Development, Maximize Open Space: Promote biodiversity by providing a high ratio of open space to development footprint.
SSc6.1	Stormwater Design, Quantity Control: Limit disruption of natural hydrology by reducing impervious cover, increasing on-site infiltration, reducing or eliminating pollution from stormwater runoff, and eliminating contaminants.
SSc6.2	Stormwater Design, Quality Control: Limit disruption and pollution of natural water flows by managing stormwater runoff.
SSc7.1	Heat Island Effect, Non-Roof: Reduce heat islands to minimize impacts on microclimates and human and wildlife habitats.
SSc7.2	Heat Island Effect, Roof: Reduce heat islands to minimize impacts on microclimates and human and wildlife habitats.
SSc8	Light Pollution Reduction: Minimize light trespass from the building and site, reduce sky-glow to increase night sky access, improve nighttime visibility through glare reduction, and reduce development impact from lighting on nocturnal environments.

to the impacts on the natural environment from forces imposed by the built environment. The IgCC, much like the IECC, is a code that regulates buildings primarily from a public welfare perspective. The IgCC is uniquely formatted not only to require the implementation of environmentally related best practices, but to encourage practices that are difficult to mandate, as well as to offer customization to jurisdictions, all in the name of reducing the negative impact of the built environment on the natural environment.

Chapter 4 of the IgCC contains requirements for site development that are intended to protect, restore, and enhance the natural features of the site (see Table 5-6).

Table 5-4
Green Globes Site Assessment by Project Stage

Pre-Design Stage
• Identification of an appropriate area for development
• Objective to respond to the site's microclimate and ecology
• Objective to preserve the site's watershed and groundwater and minimize stormwater run-off
• Objective to enhance or restore the local ecosystem
Schematic Design Stage
• Analysis of development area
• Development of strategies to minimize ecological impact
• Integration and enhancement of watershed features
• Strategies to enhance site ecology
Design Development Stage
• Development of strategies based on site analysis data
• Development of strategies to minimize ecological impact
• Integration and enhancement of watershed features
• Development of strategies to enhance site ecology
Construction Documents Stage
• Development area
• Minimization of ecological impact
Contracting and Construction
• Preservation of site watershed and groundwater
• Enhancement and preservation of local ecology

ASHRAE 189.1 Standard for the Design of High-Performance Green Buildings

In addition to the site sustainability provisions discussed in Chapter 4, ASHRAE 189.1 has mandatory requirements to mitigate urban heat island effect and minimize site light pollution. Prescriptive and performance options contain provision for pervious surfaces, protecting and restoring native and biodiverse vegetation and managing on-site storm water through reuse, infiltration or evapotranspiration.

The intent of 189.1 urban heat island mitigation section is to minimize the negative effect of heat-absorbing materials used for site hardscape, walls and roofs. At least 50% of the site hardscape must be provided by any one or combination of strategies involving shade provided by vegetation, structures and/or paving materials with a solar reflective index (SRI) of at least 29. The shade coverage on hardscape must be based on the arithmetic mean of the shade coverage calculated at 10 a.m., noon, and 3 p.m. on summer solstice. At least 30% of east and west above grade walls must be shaded from grade level to a height of 20 feet by any one or combination of strategies involving vegetation, building projections, architectural screening elements, existing buildings and/or topographical land features such as hillsides. The shade coverage calculations must be based on summer solstice

Table 5-5
LEED for Homes: Sustainable Sites

Credit	Prerequisites and Requirements
Site Stewardship: Minimize long-term environmental damage to the building lot during the construction process	
SSp1.1	Erosion controls during construction
SSc1.2	Minimize disturbed area of site
Landscaping: Design landscape features to avoid invasive species and minimize demand for water and synthetic chemicals	
SSp2.1	No Invasive Plants
SSc2.2	Basic landscape design
SSc2.3	Limit conventional turf
SSc2.4	Drought tolerant plants
SSc2.5	Reduce overall irrigation demand by at least 20%
Local Heat Island Effects: Design landscape features to reduce heat island effects	
SSc3	Reduce local heat island effects
Surface Water Management: Design site features to minimize erosion and run-off from the home site	
SSc4.1	Permeable lot
SSc4.2	Permanent erosion controls
SSc4.3	Management of run-off from roof
Nontoxic Pest Control: Design home features to minimize the need for toxic pesticides for control of insects, rodents and other pests	
SSc5	Pest control alternatives
Compact Development: Make use of compact development patterns to conserve land and promote community livability, transportation efficiency and walkability	
SSc6.1	Moderate density
SSc6.2	High density
SSc6.3	Very high density

at 10 a.m. for eastern exposed walls and 3 p.m. for western exposed walls. In Climate Zones 1, 2, and 3, at least 75% of the roof surface must comply with one or a combination of strategies involving a minimum solar reflectance index (SRI) of 78 for a low-slope roof ($\leq$ 2:12), and an SRI of 29 for a steep slope droof (>2:12), vegetated roofing, roof-covered solar energy systems, or a roof complying with ENERGY STAR criteria (see Table 5-2).

Under the prescriptive option, at least 40% of the site must incorporate any one or combination of strategies involving native vegetation, vegetated roofs, porous pavers, permeable pavement or open-graded aggregate. A number of exceptions are based on the percentage of rainfall that is captured and reused for site orbuilding water use. There is also an exception for locations with an average annual rainfall of less than 10 in. In regards to site vegetation, a minimum of 20% of a site must consist of local native plants or adapted plants based on predevelopment site conditions. A minimum of 60% of such vegetated area must consist of biodiverse plantings. The performance option provides an alternative approach to demonstrate compliance with requirements for on-site retention, collection and/or reuse of rainfall. A percentage of the average annual rainfall on the development

Table 5-6
International Green Construction Code (Public Version 2)

IgCC Chapter 4: Site Development and Land Use

Section 402: Preservation of Natural resources

402.2.1 Floodplains. Buildings and building site improvements shall not be located within a floodplain.

402.2.2 Surface water protection. Buildings and building site improvements shall not be located within a buffer around a water body or within a buffer around a water body, as defined as the ordinary high-water mark of seas, lakes, rivers, streams and tributaries which support or could support fish, recreation or industrial use.

402.2.3 Conservation area. Site disturbance or development of land within 50 feet of any designated conservation areas shall not be permitted.

402.2.4 Park land. Site disturbance of development of land located within a public park shall not be permitted.

402.2.5 Agricultural land. Buildings and associated site improvements shall not be located on land zoned for agricultural purposes.

402.2.6 Greenfield sites. Site disturbance or development shall not be permitted on greenfield sites.

402.3.1 Predesign site inventory and assessment. An inventory and assessment of the natural resources and baseline conditions of the building site shall be submitted.

402.3.2 Stormwater management. Stormwater management systems, including, but not limited to, infiltration, evapo-transpiration; rainwater harvesting and runoff reuse, shall be provided and maintained on the building site. Stormwater management systems shall address the increase in runoff that would occur resulting from development on the building site.

402.3.3.1 Water for outdoor landscape irrigation. Water used for outdoor landscape irrigation shall be non-potable.

402.3.4 Outdoor ornamental fountains and water features. Outdoor ornamental fountains and other water features constructed or installed on building site shall be supplied with either municipally reclaimed or collected rainwater.

402.3.5 Management of vegetation, soils and erosion control. During construction on a building site, vegetation and soils shall be protected, selected and reused.

402.3.6 Building site waste management plan. A building site waste management plan shall be developed and implemented to recycle or salvage not less than 75 percent of land-clearing debris and excavated soils in accordance with specified provisions. Land-clearing debris includes rocks, trees, stumps and associated vegetation.

Section 404: Heat Island Mitigation

404.2 Site Hardscape. Not less than 50 percent of site hardscape shall be provided with one or any combination of the following options: 1) hardscape materials with a minimum initial SRI of 29; 2) shading structures; 3) shade trees; 4) pervious paving, open-grid pavers, and/or open-graded aggregate (stabilized decomposed granite).

404.3 Roof coverings. Not less than 75 percent of roof surfaces of buildings shall be provided with one or any combination of the following options: 1) roof surfaces with minimum solar reflectance/thermal emittance or SRI value per Table 404.3.1 or 2) installation of a vegetative roof.

Section 406: Detailed Site Development Requirements

406.2 Non-potable water systems for irrigation systems. Non-potable water systems used for irrigation shall comply with the graywater, municipal reclaimed water and collected rainwater provisions of this section.

406.3 Subsurface graywater irrigation systems. Gravity subsurface gray water irrigation systems shall be designed and installed in accordance with section.

406.4 Vegetation and soil protection. Vegetation and soil protection plans shall comply with the identification and protection requirements of this section.

406.5 Soil reuse and restoration. Soils that are reused and restored on building site shall comply with preparation and restoration requirements of this section.

(Continued)

Table 5-6 (*Continued*)

406.6 Landscape, soil and water quality protection plan. A landscape, soil and water quality protection plan shall comply with erosion, sedimentation, pollutant control and maintenance protocol requirements of this section.

406.7 Vegetative roofs. Vegetative roofs shall comply with specified provisions for plant selection, engineered soil medium and roof membrane protection in accordance with this section.

footprint (on-site impervious surfaces) must be managed through infiltration, re-use or evapotranspiration (ET) based on whether the site is on an existing building site, greyfield, brownfield or greenfield site.

REGIONAL AND LOCAL PLANNING CONSIDERATIONS

Every region of the country has its own set of environmental characteristics, including constraints and unique issues. Each building site also has its special conditions involving topography, hydrology, and vegetation that connect to a larger ecosystem. The goal of sustainable site development is to preserve the natural resources of the site in the context of the region while minimizing the environmental impact of the development. This involves preservation of native vegetation and open space, erosion control, protection of topsoil, restoration of habitats, minimizing heat-island effect, and providing for pedestrian connectivity to the surrounding community. This requires interfacing with local planning and zoning regulations regarding uses, building setbacks, density, building heights, open space, street access, parking, and utility infrastructure. Planning issues are the most difficult to standardize from community to community. There is no "one-size-fits-all" solution.

WATER EFFICIENCY

INDOOR WATER-EFFICIENCY MEASURES

Water use in buildings accounts for about 12% of the freshwater withdrawals. Residential water use accounts for more than half of the publicly supplied water in the United States. In the average American home, 70% of total household water consumption is from indoor plumbing use and 30% from outdoor use. During summer months in arid climates, these percentages are often reversed.[1]

Water efficiency measures are intended to:

- Reduce or eliminate potable water use for landscape irrigation
- Reduce demand of potable water for indoor use
- Encourage on-site reuse of graywater and collected rainwater
- Reduce demand on publicly supplied water systems
- Reduce wastewater discharge and burden on wastewater treatment facilities
- Preserve the quality of water in aquifers, lakes, rivers, streams, and reservoirs

Water also contains embodied energy. Energy is used to treat and convey water to and within buildings, and significant amounts of energy are used to heat water in buildings. Thus, efficient heating and distribution of water within buildings also conserves energy.

Many of the features discussed in Chapter 5, including stormwater management, landscaping, rainwater collection, and vegetated roofs, support water-efficiency measures. This chapter primarily addresses indoor water-efficiency measures.

Defining Green

Rainwater Harvesting

Rainwater collection system extending between the collection surface and the storage tank that conveys collected rainwater for use in landscape irrigation and other non-potable uses where permitted by local building and health codes.

DESIGN WITH LESS IMPACT ON NATURAL WATER SOURCES

As described in Chapter 4, the hydrologic cycle is an enormous cycling of water through the earth's atmosphere, oceans, land, and biosphere that shapes the world's climate, supports plant growth, and makes life possible. Water is vital to the survival of everything on the planet and is limited in supply. The earth might seem like it has abundant water, but in fact less than 1% is available for human use. The rest is either salt water found in oceans, fresh water frozen in the polar ice caps, or too inaccessible for use. While the population and the demand

[1]EPA WaterSense Resource Manual (February, 2010).

on freshwater resources are increasing, supply remains constant. Managing water is a growing concern in the United States. Communities across the country are facing challenges regarding water supply and waste water infrastructure. As the map in Figure 6-1 shows, many of the states that have projected population growth increases also have higher per capita water use.

Figure 6-1

Projected water usage.

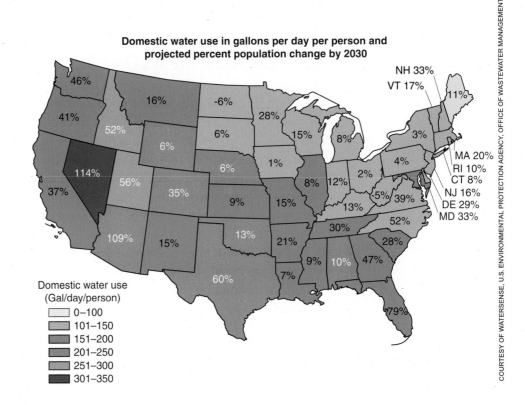

Domestic water use in gallons per day per person and projected percent population change by 2030

Water Shortage

Water shortage is an increasing national and global problem. The U.S. population grew by 89% between 1950 and 2000. During that time, water usage increased 200%[2] (see Figure 6-2). The average American uses the equivalent of 100 gallons of water a day, which is double the average in Europe. In addition, through an aging infrastructure, millions of gallons are lost daily due to leaks in water-supply piping.

The efficient use of water reduces stress on the urban water infrastructure. This equates to reduced energy usage, as large amounts of energy are used to treat and transport water. Water efficiency also means the avoidance of water resource disputes. During the summer of 2009, a federal judge threatened to cut off the city of Atlanta's water supply. The U.S. Court of Appeals sought to end a long tri-state water war between Georgia, Alabama, and Florida. The judge gave the state of Georgia three years to end its reliance on Lake Lanier, a 38,000-acre federal reservoir north of Atlanta that is the main water source for the city's more than 3 million residents. However, in June of 2011, a federal appeals panel overturned the lower court ruling

---

[2]U.S. Geological Survey, *Estimated Use of Water in the United States in 2000*, USGS Circular 1268 (2005).

Figure 6-2

Water usage between 1950 and 2000.

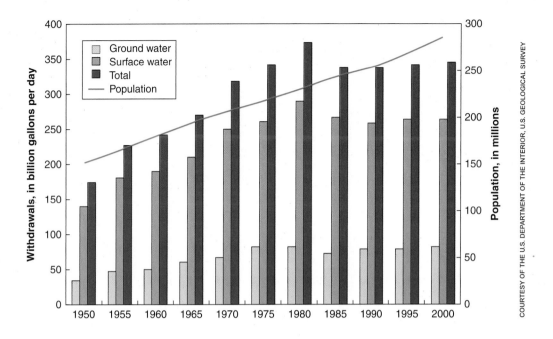

COURTESY OF THE U.S. DEPARTMENT OF THE INTERIOR, U.S. GEOLOGICAL SURVEY

and said the 3 million people around metro Atlanta can continue to draw water from Lake Lanier.[3]

Local water-supply issues vary greatly across the country because water is a regional issue. State regulations usually govern the reuse of water and its application with respect to graywater, rainwater collection, and reclaimed water. Water consumption can be reduced through improved design and smart technologies that conserve water and reduce demand on water supply. Indoor potable water consumption can be reduced by using high-efficiency plumbing fixtures and by reducing the volume of water in the distribution system between service water heating equipment and plumbing fixtures.

Outdoor landscape water can account for over half of potable water consumption in hot and arid regions of the country[4] (see Figure 6-3). Irrigation needs can be reduced through the use of native plants and drought-tolerant landscape design (see Figure 6-4). Irrigation can be supplied by non-potable water sources, including graywater and municipal reclaimed water. Reducing our demand on potable water reduces energy and chemical use at municipal

Defining Green

Embodied Energy of Water

Energy is needed to source, treat, and distribute municipal water. The amount of energy and associated emissions vary depending on the utility's water sources, the distances transported, and the type of water treatment applied. Employing water-efficiency strategies at the building site can greatly reduce the environmental impacts of water use.

[3]Atlanta Regional Commission, *Tri-State Water Wars* - http://www.atlantaregional.com/environment/tri-state-water-wars

[4]EPA WaterSense Resource Manual (February 2010)

Figure 6-3

Irrigation for turf in a hot and dry desert environment.

COURTESY OF ANTHONY FLOYD

Figure 6-4

A xeriscape landscape conserves water while providing for an esthetically pleasing environment at ground level and vegetated terrace levels.

COURTESY OF ANTHONY FLOYD

water treatment facilities. As such, wastewater reuse lessens the energy demand of wastewater treatment plants and in turn reduces the amount of wastewater released into environmentally stressed bodies of water, including streams and rivers.

Water Conservation and Efficiency

Water conservation is difficult without water efficiency. Conservation without efficiency ignores the larger context of system design and operations. Efficiency means using resources responsibly, planning for usage, and employing smart technologies that make it easier for consumers to do the right thing without sacrificing expected outcomes.

Inefficient water conserving products have negatively impacted the plumbing industry, making it harder to break through to a skeptical public about the broader benefits of using less water. The plumbing industry was negatively impacted when federal legislation mandating low-flow toilets in the early 1990s preceded development of efficient technology. Manufacturers have long since solved those problems, but many consumers remember.[5]

The U.S. EP Act of 1992 (Table 6-1) made "low-flow" toilets a minimum requirement for new homes and home remodels beginning in 1994. Problems were discovered with some of the early low-flow toilets. Consumers complained about having to flush twice to clean the bowl, bringing water usage back up to 3.5 gallons per flush (gpf); stoppages; as well as water spots that made it harder to keep the toilet bowl clean and sanitary.

The objective of the U.S. EP Act of 1992 was water conservation, but the objective of plumbing fixture designers and manufacturers was water efficiency. With improved technology, and by the late 1990s consumer reports found several affordable low-flow toilets that worked very well.[6] Using advanced hydraulic modeling techniques, engineers were able to change the water flow dynamics of the toilet trap to eliminate waste. Newer finishes fired into the chinaware gave more power to less water, actually improving bowl-cleansing ability over the old 3.5 gpf toilets.

Plumbing manufacturers were soon able to introduce high-efficiency models that use even less water, yet achieved the same satisfactory performance at a reasonable

Table 6-1
Energy Policy Act of 1992

Fixture	Flow Requirement
Water closets (gallons per flush)	1.6
Urinals (gallons per flush)	1.0
Showerhead (gallons per minute)	2.5
Faucets (gallons per minute)	2.2
Metering faucets (gallons per cycle)	0.25

COURTESY OF WATERSENSE, U.S. ENVIRONMENTAL PROTECTION AGENCY, OFFICE OF WASTEWATER MANAGEMENT

[5]Safe Plumbing, available at http://safeplumbing.org/water-efficiency

[6]May 1998 issue of *Consumer Reports* magazine.

price for consumers. High-efficiency toilets (1.28 gpf) reduce water use and the strain on sanitary systems (see Figure 6-5). The use of high-efficiency plumbing fixtures can delay or eliminate the need for developing new or expanded municipal water and wastewater treatment facilities, saving consumers and taxpayers millions of dollars. Utilities have invested hundreds of millions of ratepayer dollars in water conservation programs that rely on water-efficient plumbing fixtures and appliances.

With water and wastewater infrastructure costs running millions of dollars each year, communities can rely on the steady water savings derived from fixtures that use less water. Even where water is not scarce, efficient plumbing fixtures help consumers and communities reduce the strain on aging water utility infrastructures. While consumers save on energy, water, and wastewater costs, communities save on infrastructure upgrades.[7]

Figure 6-5

Dual flush, high efficiency toilet.

COURTESY OF ANTHONY FLOYD

Wastewater Conveyance and Water Efficiency

One of the unintended consequences of the efficient use of water is less wastewater flowing down the drain lines. Dry drains are the result of high-efficiency plumbing fixtures, water-conserving appliances, and graywater reuse that reduce the amount of water discharged into the waste drain system. With the enactment of the EP Act of 1992, all water closets (toilets) were required to flush no more 1.6 gpf, effective January 1, 1994, for residential models and January 1, 1997, for all other models. At that time, concern for drain line transport efficacy was expressed by the plumbing industry. However, early research only focused on the

[7]Safe Plumbing, available at http://safeplumbing.org/water-efficiency

flush efficacy of the various water closets. Drain line transport problems were largely attributed to faulty sanitary drain lines.[8]

With the advent of high-efficiency toilets (HETs) with a maximum average of 1.28 gpf, some water closet manufacturers are now voluntarily offering models that flush at 1.0 and 0.8 gpf. This has raised the debate of drain line transport efficacy. Many plumbing experts are concerned that we are at or approaching a "tipping point" where a significant number of sanitary waste systems will be affected by drain line transport problems, especially in larger commercial systems that have an isolated long horizontal drain line with a single fixture drain.[9] Municipal wastewater departments are not as concerned because municipal sewer lines accumulatively collect wastewater that generates greater flow rates downstream along the sewer system.

Newer technologies, such as zero-water and high-efficiency urinals (HEUs) with a maximum 0.5 gpf (see Figure 6-6), lower-flow-rate faucets, and increasingly efficient water-consuming

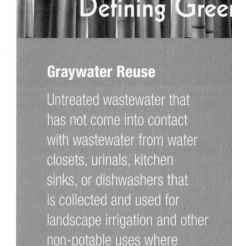

Defining Green

Graywater Reuse

Untreated wastewater that has not come into contact with wastewater from water closets, urinals, kitchen sinks, or dishwashers that is collected and used for landscape irrigation and other non-potable uses where permitted by local building and health codes. Graywater includes but is not limited to wastewater from bathtubs, showers, lavatories, clothes washers, and laundry trays.

Figure 6-6

Waterless urinal (max. 0.5 gpf).

COURTESY OF ANTHONY FLOYD

[8, 9]George, Ron, *Code Update: Water Conservation Research and Legislative News, Part I*, Plumbing Engineer, 8/11

Figure 6-7

Graywater system schematic.

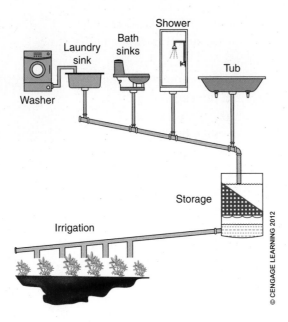

appliances will further reduce the amount of water discharged into drain lines. Graywater systems (see Figure 6-7), which collect discharged water from lavatory basins, clothes washers, bathtubs, and shower fixtures for reuse as irrigation or toilet flushing, are another water-efficiency approach that reduces wastewater in drainage systems. Research is needed to determine the minimum wastewater flow rate needed to effectively transport solid waste down the drain per pipe size and slope.[10]

Non-Potable Water Use

Effective means of reducing the use of potable water in buildings include the effective use of non-potable water where it is appropriate and is permitted by codes within the jurisdiction. Non-potable water is generally considered to be water that is not suitable for drinking purposes, as determined by the jurisdiction. Non-potable water includes the reuse of water within buildings, as facilitated by graywater, rainwater catchment and reclaimed water systems.

Many green codes, standards, and rating systems encourage the use of non-potable water, and discourage or prohibit the use of potable water, for toilet and urinal flushing, water-supplied trap primers, industrial uses including carwash facilities, and irrigation. Water quality for each end use must be in accordance with the laws, rules and ordinances of the jurisdiction.

The International Plumbing Code (IPC) contains an appendix that addresses the design, construction and installation of graywater systems. Appendices in the IPC are applicable only where specifically adopted by the local jurisdiction of the project location. The IgCC contains detailed provisions for the design, construction and installation of graywater, rainwater catchment and reclaimed water systems.

[10]George, Ron, Code Update: Water Conservation Research and Legislative News, Part I, Plumbing Engineer, 8/11

The IgCC's graywater provisions permit water to be collected from bathtubs, showers, lavatories, clothes washers, laundry trays and, where approved by the code official, swimming pool backwash, air conditioner condensate, rainwater, cooling tower blow-down, foundation drainwater, steam system condensate, fluid cooler discharge, food steamer discharge, combination oven discharge, industrial process water and fire pump test water.

All non-potable water systems must be separated from the potable water system and usually requires labeling and color coding of piping. Furthermore, some fixtures and equipment may not be designed to utilize non-potable water, and their service life may be compromised, warranties may be voided and increased maintenance may be required. On the other hand, newly developing technologies may soon make such water technologies a reality within buildings.

Water Use Associated with HVAC Systems and Equipment

Green codes, standards and ratings systems often limit, discourage or prohibit the use of potable water in association with HVAC systems and equipment, including water associated with hydronic heating and cooling systems, ground-source heat pumps, humidification systems, condensate coolers, heat exchangers, cooling towers, evaporative condensers, fluid coolers and humidifiers. The collection of water discharged from some such systems and equipment may also be required or encouraged to use for irrigation, outdoor water features or toilet flushing.

Metering

Metering and sub-metering of water used by plumbing and mechanical systems and subsystems can provide building owners and facility managers with valuable information regarding the performance of those systems and subsystems. Discrepancies in the water used by systems and equipment as compared to similar previous periods can be used as warning signs that prompt searches for leaks and inefficiencies associated with water distribution systems. Metering is a simple and effective tool that provides instant feedback and encourages building owners to operate and maintain systems that were designed to conserve water in the most efficient manner over the life of the building. More water use typically means more cost, and that is usually a very effective incentive to be efficient.

WATER RESOURCE TOOLS AND GUIDELINES

EPA WaterSense Program

The WaterSense label is a symbol to consumers that a product uses less water and performs as well as or better than conventional models. Before a product can earn the WaterSense label, it must be independently tested and certified to meet Environmental Protection Agency (EPA) criteria for efficiency and performance. The label can be found on tank-type toilets, residential bathroom faucets, faucet accessories, and flushing urinals that use at least 20% less water than the standard, and more products are being considered for the label every year (see Figure 6-8).

The WaterSense new homes program is designed to actively promote the transformation of the mainstream homebuilding industry toward increased water efficiency. The WaterSense Single-Family New Home Specification provides national consistency in defining the features of a water-efficient new home and enables builders anywhere in the country to obtain a WaterSense label on their homes. Builders and other users can go beyond the criteria in the specification, which identifies the minimum requirements for a water-efficient new home.

WaterSense-labeled new homes combine WaterSense-labeled products with other water-efficient fixtures, systems, and practices to reduce the amount of water used by approximately 20% compared to typical new homes (see Table 6-2). In

Figure 6-8

EPA WaterSense logo.

COURTESY OF WATERSENSE, U.S. ENVIRONMENTAL PROTECTION AGENCY, OFFICE OF WASTEWATER MANAGEMENT

addition to WaterSense-labeled toilets and faucets, these new homes include dishwashers and clothes washers with the ENERGY STAR label, if such appliances are installed when the home is built. WaterSense-labeled new homes incorporate an efficient hot water distribution system that decreases the amount of time it takes for hot water to reach the faucet or shower, as waiting for hot water can waste thousands of gallons of water each year. A maximum water service pressure in the house is set in each WaterSense-labeled home to reduce the maximum water flow from fixtures and the likelihood of leaking pipes and hoses.

Outside the home, builders are required to meet minimum landscaping requirements using either a performance-based or prescriptive-based approach to determine the allotment of turfgrass allowed in the front yard. If an landscape irrigation system is installed, WaterSense outdoor water criteria require landscape water efficiency through a combination of appropriate landscape and efficient irrigation system design. Certain landscapes in many regions of the country may not need supplemental water. To reduce the use of potable water, builders can use alternative sources of water, such as rainwater, graywater, or municipally reclaimed water. The use of non-potable water is not intended to substitute for water efficiency, but rather to supplement existing efficiency measures.

Table 6-2
WaterSense Expected Water Savings

Indoor Features	Standard Water Use	Standard Use (gal/day/household)	WaterSense Criteria	Expected Use (gal/day/household)	Expected Water Savings (gal/day/household)
Toilets	1.6 gpf	21.0	1.28 gpf	16.8	4.2 (20%)
Bathroom faucets	2.2 gpm	29.1	1.5 gpm	27.6	1.5 (4.8%)
Showerheads	2.5 gpm	25.4	2.5 gpm	25.4	0 (0%)
Hot water delivery systems	~10 gallons per day per household wasted		Assumes 20% water savings for improved design	8.0	2.0 (20%)
Dishwashers	8.6 gallons per load (6 gallons per cycle)	2.7	5.8 gallons per load (4 gallons per cycle)	1.8	0.9 (33%)
Clothes washers	39.6 gallons per load (12 gallons per cycle per cubic foot)	39.9	24 gallons per load (6 gallons per cycle per cubic foot)	22.0	17.9 (45%)
Total Indoor		128.1		101.6	26.5 (20.7% savings)

LEED-NC

LEED-NC is a rating system composed primarily of elective provisions that are assigned points. LEED-NC contains few mandatory provisions. Its mandatory provisions are its "prerequisites." The intent of LEED-NC's Water Efficiency section is to reduce the use of potable water and the generation of wastewater while increasing the local aquifer recharge. There is one prerequisite and 34 optional credits for a total of 10 available points (Table 6-3).

Green Globes

Green Globes is a rating system containing elective provisions that are assigned point values. Green Globes does not contain mandatory requirements. The Green Globes Water category assessment consists of online survey questions organized under the following six stages of the design construction process: (1) Pre-design, (2) Schematic Design, (3) Design Development, (4) Construction Documents, (5) Contracting/Construction, and (6) Commissioning. Each stage assesses water conservation priorities with the objective of reducing environmental impacts (Table 6-4).

Table 6-3
LEED for New Construction: Water Efficiency

Credit	Prerequisite and Credit Intents
WEp1	Water Use Reduction: Employ strategies that in aggregate use 20% less water than the water use baseline calculated for the building (not including irrigation).
WEc1	Water Efficient Landscaping: Reduce potable water consumption for irrigation by 50% from a calculated midsummer baseline case OR Use non-potable source of water or install landscaping that does not require a permanent irrigation system.
WEc2	Innovative Wastewater Technologies: Reduce potable water use for building sewage conveyance by 50% through the use of water-conserving fixtures (e.g., water closets, urinals) or non-potable water (e.g., rainwater, graywater or municipally reclaimed water) OR Treat 50% of wastewater on-site to tertiary standards. Treated water must be infiltrated or used on-site.
WEc3	Water Use Reduction: Employ strategies that in aggregate use 30%, 35% or 40% less water than the water use baseline calculated for the building (not including irrigation).

Table 6-4
Green Globes Water Assessment by Project Stage

Pre-Design Stage

- Objective to establish a water target
- Objective to minimize the demand for potable water
- Objective to minimize the need for the off-site treatment of water

Schematic Design Stage

- Meeting a water performance target
- Water conserving strategies
- Strategies to reduce off-site treatment of water

Design Development Stage

- Water performance target
- Water conserving strategies
- Strategies to minimize off-site treatment of water

Construction Documents Stage

- Water performance
- Water-conserving features
- Minimization of off-site treatment of water

Contracting and Construction

- Water-conservation

Commissioning

- Commissioning for minimization of potable water consumption
- Commissioning of water treatment systems

LEED for Homes

LEED for Homes is a rating system composed primarily of elective provisions that are assigned points. LEED-NC contains few mandatory provisions. Its mandatory provisions are its "prerequisites." The intent of LEED for Home's Water Efficiency section is to reduce the use of potable water and the generation of wastewater while increasing the local aquifer recharge. There are 4 optional credits offering 15 available points in the water section, and a minimum of 3 points is required (Table 6-5).

CODES AND STANDARDS

GBI-01 Green Building Assessment Protocol for Commercial Buildings

GBI-01 is a rating system. It is also an ANSI standard. Chapter 9 of GBI-01 regulates water. GBI-01 requires that at least 26 percent of the 130 points available in the water efficiency provisions in Chapter 9 be satisfied. In addition, Table 9.1.1 of GBI-01 requires that specific adjustments be made when calculating points related to water (see Table 6-6).

ICC 700 National Green Building Standard

ICC 700 is a rating system. It is also an ANSI standard. Chapter 8 of ICC 700 addresses water efficiency. The only mandatory requirement in Chapter 8 of ICC 700 is that, for its Gold and Emerald performance levels, either Section 801.6 or 802.2 must be complied with. The standard does, however, require increased points in water at each of its performance levels. A minimum of 14 points at the Bronze, 26 points at the Silver, 41 points at the Gold and 60 points at the Emerald performance level must be acquired from the water efficiency provisions of ICC 700 Chapter 8. See Table 6-7 for a summary of these provisions.

Table 6-5
LEED for Homes: Water Efficiency

Credit	Requirements
WEc1.1	Rainwater harvesting
WEc1.2	Graywater reuse system
WEc1.3	Use of municipal reclaimed water
WEc2.1	High efficiency irrigation system
WEc2.2	Third party inspection
WEc2.3	Reduce overall irrigation demand by at least 45%
WEc3.1	High-efficiency plumbing fixtures
WEc3.2	Very high-efficiency plumbing fixtures

Table 6-6
GBI-01 Green Building Assessment Protocol for Commercial Buildings

Chapter 9: Water

Plumbing Fixtures, Fittings, Appliances, and Equipment

- 6 to 24 points awarded where projected water consumption is reduced at least 25 to 40% as calculated with the Green Globes Water Consumption Calculator.
- 18 points awarded where all plumbing fixtures and fittings are in WaterSense compliant or, for types of fixtures and fittings not addressed by WaterSense, in accordance with the Green Globes Water Consumption Calculator.
- 2 points awarded, up to a total maximum of 4 points, for each EnergyStar residential clothes washer with a max water factor of 6 gallon/ft^3 per cycle or EnergyStar residential dishwater with a max water factor of 5.8 gal/ft^3.

Cooling Towers

- 6 points awarded where water total hardness, discharge conductivity is in accordance with 9.3.1.1.
- 6 points where a cooling tower water treatment program in accordance with 9.3.1.2 is established.
- 1 point, to a maximum of 4 total points, awarded where at least 20 to 51% of annual heat rejected is dissipates with sensible (dry) cooling or no wet cooling is provided.
- 2 points awarded where cooling towers are equipped with drift eliminators in accordance with 9.3.1.4.

Boilers and Water Heaters

- 3 points awarded where boilers or water heaters are installed with feed makeup water, boilers over 50 bhp are provided with condensate return systems, boilers are fitted with conductivity controller and steam boilers are provided with conductivity meters.

Commercial Food Service Operations

- 3 points where once-through water-cooled equipment is not used. N/A where there no commercial food service operations in the building.
- 2 points where garbage disposals are not water-fed.
- 2 points where ice machines are ENERGY STAR compliant.
- 1 point awarded where combination ovens use no more than 10 gph, or 4 points if use no more than 4 gph.
- 1 point awarded where pre-rinse spray valves are present and comply with the US Energy Policy Act of 2005.
- 1 point awarded where not more than 2.0 gallons of water per hour are used by boilerless/connectionless food steamers.
- 1 point awarded where dishwashers are ENERGY STAR compliant.

Medical/Dental and Laboratory Facilities

- 2 to 4 points awarded where steam sterilizers comply with 9.6.1.
- 3 points awarded where non-potable water is used for once through cooling of laboratory or medical equipment.
- 2 points awarded where all vacuum systems are of the dry type.
- 1 point awarded where digital imaging technology is used for x-rays or film processing of x-rays more 150 mm in any dimension are provided with water recycling units.
- 1 point awarded where wet scrubbers are provided with water recirculation systems.

Commercial/Institutional Laundry Operations

- 1 to 10 points awarded where the water factor for clothes washers is less than 19.4 to 7.5 gal/ft^3.

Special Water Features (swimming pools, spas, ornamental fountains and water playscapes)

- 1 point awarded where water is filtered and re-circulated for reuse within a system.
- 1 point awarded where meters are installed for potable water makeup lines for all special water features.
- 1 point awarded where alternative water sources are used makeup water.

Table 6-6 (*Continued*)

Water Treatment

- 2 points awarded where filtration systems are equipped with pressure drop buages in accordance with 9.9.1.1.
- 2 points awarded where reverse osmosis systems are in accordance with 9.9.1.2.
- 1 point awarded where water softeners are provided with recharge controllers in accordance with 9.9.1.3.

Alternate Sources of Water

- 2 to 15 points awarded where at least 10 to 60% of potable is replaced by alternate sources.

Metering

- 4 points awarded where water meters and submeters and a meter data management system is provided in accordance with 9.11.1.1.
- 2 points awarded where makeup meters are prohibited for chilled or hot water loops.

Table 6-7
ICC 700 National Green Building Standard

Chapter 8: Water Efficiency

Indoor and outdoor water use

Hot Water Distribution System Design

Encourages reduced hot water use addressing hot water supply piping system design and installation as follows:

1) 2 points awarded where runs to kitchen and bath fixtures are not more than 40 ft in length from the water heater

2) 3 points awarded where runs to kitchen and bath fixtures are not more than 30 ft in length from the water heater

3) Allows points for one of the following:

    a) 6 points awarded where structure-type plumbing with demand-controlled loops are used and the volume of water in the pipe and fittings downstream of the trunk line is limited to 4 cups

    b) 6 points awarded where piping a parallel piping system is not more than 15 feet in length from fixture fittings to hot water heater and the water volume is limited to 8 cups.

    c) 8 points awarded where each pipe run between water heater and fixture fittings in a central core system is limited to 6 cups.

4) 1 point awarded where pipe runs between hot water heater and fittings and either

    a) a tankless water heater is provided at point of use and is supplied by a solar-assisted or cold water system, or

    b) the hot water system is of the on-demand recirculation type

Appliances

Points for ENERGY STAR water-conserving appliances as follows:

1) 2 for dishwasher

2) 8 for washing machine

3) 12 for washing machine where it has a water factor of not more than 6.0.

Disposers

1 point awarded where the kitchen sink is provided with a food waste disposer.

Shower Heads

1) 1 point awarded, up to 3 max, for each shower compartment with a flow rate of 1.6 to 2.5 gpm

(Continues)

Table 6-7 (*Continued*)

> 2) Either
>
>   a) 1 additional point awarded where all shower head flow rates are 2.0 to less than 2.5 gpm, or
>
>   b) 2 additional points awarded where all shower head flow rates are 1.6 to less than 2.0 gpm

Lavatories

> 1) 1 point, up to a max of 3 total points, awarded for each bathroom where all lavatory faucets have a maximum flow rate of 1.5 gpm, and an additional 2 points where all lavatory faucets in the dwelling unit meet this requirement,
>
> 2) 1 point, up to a max of 3 total points, awarded for each faucet that is equipped with a motion sensor or self-closing valve equipped, is of the metering or self-closing valve type or or is pedal-activated.

Water Closets and Urinals

> Compliance required and points awarded as follows:
>
> 1) Compliance with 802.2 or 801.6 is required for the Gold and Emerald performance levels
>
> 2) 6 points are awarded, up to a maximum of 18, for each water closet with a maximum flush volume of 1.28 gallons
>
> 3) 4 points awarded where all urinals have a maximum flush volume of 0,5 gallons.
>
> 4) 6 additional points are awarded when all urinal and water closets in the dwelling unit are in accordance with items 2 and 3.

Irrigation Systems

> Awards 2 to a maximum of 10 total points for low-volume irrigation systems of various types, and an additional 3 points where the system is in installed in accordance with and designed by a professional in accordance with EPA WaterSense requirements.

Rainwater Collection

> Awards 6 points where rainwater is collected and reused and additional 2 points where it is distributed using gravity or a renewable energy source.

Innovative Practices

> 1) 4 points awarded for each water closet that uses reclaimed or recycled water for flushing
>
> 2) 10 points awarded where reclaimed or on-site recycled water is used for irrigation purposes
>
> 3) 8 points awarded for each waterless or composting urinal or toilet, and an additional 8 points where all toilets are in accordance with 802.2(2), up to a total maximum of 24 points.

ASHRAE 189.1 Standard for the Design of High-Performance Green Buildings

The water-use-efficiency section of ASHRAE 189.1 addresses water waste in buildings and on surrounding landscape. It also establishes provisions to monitor and manage water consumption. ASHRAE 189.1 includes a set of mandatory requirements that must be met for all projects and an option for either a prescriptive or performance set of requirements to demonstrate compliance. The prescriptive option outlines a series of specific water-saving measures, and the performance option allows a project's overall water use to be calculated and compared to a maximum baseline level.

Mandatory Building Water-Use Reduction

Mandatory building water-use reduction includes a 20% water use reduction (Table 6-8) using the EPAct of 1992 as a benchmark. In some cases the fixtures must be WaterSense certified. All appliances must be ENERGY STAR rated.

The mandatory section also contains water-efficiency requirements for HVAC equipment. "Once-through" cooling systems are prohibited. Because once-through cooling systems do not recirculate the cooling water, they can use thousands of gallons of water per day. Condensate from steam systems and large air-conditioning units (larger than 65,000 Btu/h [19,000 W]) must be recovered and reused. Air handler condensate recovery systems can reroute the collected condensate from the air handling units (AHUs) to cooling towers to be used as makeup water, reducing the amount of potable water needed for cooling tower processes. The recovered condensate could also be used for landscape irrigation.

Mandatory Water Consumption Measurement

Both potable and reclaimed water entering the building project must be monitored or submetered where required. Measuring and monitoring water use helps to identify water-use anomalies that might occur when building and process systems break down. When connected to a real-time building management system, such events can be quickly addressed and corrected. Monitoring systems must be capable of remotely recording and storing the collected data on an hourly, daily, and monthly basis and must be equipped to alert operators to problems in real time.

Table 6-8
ASHRAE 189.1 Plumbing Fixture Requirements

Plumbing Fixture	Maximum Volume
Water Closets (Toilets)	
• Flushometer Valve Type	Single Flush: 1.28 gal (4.8 L)
• Flushometer Valve Type	Effective Dual Flush: 1.28 gal (4.8 L)
• Tank-Type	Single Flush: 1.28 gal (4.8 L) and WaterSense certified
	Effective Dual Flush: 1.28 gal (4.8 L) and WaterSense certified
• Urinals	0.5 gal (1.9 L)
Faucets	
• Public Lavatory	0.5 gpm (1.9 L/min)
• Public Metering Self-Closing	0.25 gal (1.0 L) per metering Cycle
• Residential Bathroom Lavatory Sink	1.5 gpm (5.7 L/min) and WaterSense certified
• Residential Kitchen	2.2 gpm (8.3 L/min)
Showerheads	
• Residential	2.0 gpm (7.6 L/min)
• Residential shower compartment (stall) in dwelling units and guestrooms	All shower outlets: 2.0 gpm (7.6 L/min)

ASHRAE JOURNAL JUNE 2010. © AMERICAN SOCIETY OF HEATING, REFRIGERATING AND AIR-CONDITIONING ENGINEERS, INC. WWW.ASHRAE.ORG.

Perspective and Performance Option

The perspective compliance option requires additional building water use reduction including HVAC equipment, commercial food service equipment, medical equipment and processes, and water features. The performance compliance option must demonstrate compliance by completing performance-based calculation for either site or building water use, or both.

International Green Construction Code (IgCC)

The IgCC is a code composed primarily of minimum mandatory requirements. As mentioned in previous chapters, the International Green Construction Code (IgCC) provides a comprehensive set of requirements intended to reduce the negative impact of buildings on the natural environment. Chapter 7 of the IgCC contains requirements for the efficient delivery, use, and reuse of water indoors and outdoors (see Table 6-9).

Table 6-9
International Green Construction Code (Public Version 2)

Chapter 7: Water Resource Conservation and Efficiency
Section 702: Fixtures, Fittings, Equipment, and Appliances
The aggregate *potable* water consumption of fixtures and fittings shall be 20 percent less than the reference values as specified in Section 702.1, with exceptions.   Specific conservation provisions (Sections 702.2 thru 702.19):   • Combination tub/shower valves   • Food establishment pre-rinse spray heads   • Drinking fountain controls   • Non-water urinal connections   • Appliances (clothes washers, ice makers, food steamers and dishwashers)   • Efficient hot water distribution systems   • Makeup water supply   • Food service handwashing faucets   • Spa covers, Swimming pool covers and splash troughs
Section 703: HVAC Systems and Equipment
Specific conservation provisions (Sections 703.1 thru 703.6):   • Hydronic closed systems   • Humidification systems   • Condensate coolers and tempering   • Condensate drainage recovery   • Heat exchangers   • Humidifier discharge

Table 6-9 (*Continued*)

Section 704: Water Treatment Devices and Equipment

Water softeners shall comply with the water conservation provisions of Section 704.1:

- Demand initiated regeneration
- Water consumption
- Water connections
- Efficiency and listings

Reverse osmosis water treatment systems shall comply with water conservation provisions of Section 704.2.

Section 705: Specific Water Conservation Measures

Where indoor ornamental fountains, indoor water features or permanent indoor irrigation systems are supplied by *potable water*, the *building* that contains them shall comply with one additional *project elective* from Section 710. All *potable* and non-*potable* water shall be individually *metered* in accordance with the requirements indicated in Table 705.2.1. Each *meter* shall be required to be capable of communicating water consumption data remotely.

Section 706: Non-Potable Water Requirements

Where *non-potable* water is used for a water use application, signage shall be provided per Section 706.2. *Non-potable* water for each end use application shall meet the minimum water quality requirements as established for the application by the laws, rules and ordinances applicable in the *jurisdiction*.

Section 707: Rainwater Collection and Distribution Systems

Provisions for the construction and installation of rainwater collection and conveyance systems.

Section 708: Graywater Systems

Provisions for the construction and installation of graywater reuse systems.

Section 709: Reclaimed Water Systems

Provisions for the construction and installation of systems supplying non-potable reclaimed water.

Not only does the IgCC contain mandatory requirements for the implementation of numerous water conservation and efficiency related best practices, it also provides detailed requirements to guide the design and construction of irrigation, rainwater catchment, graywater and reclaimed water systems. Unlike the permissive language typically used by guidelines and voluntary rating systems, the text of each IgCC code section is written in mandatory language that is readily enforceable and is also integrated with the requirements of the International Plumbing Code (IPC).

Table 302.1 of the IgCC allows the local jurisdiction to require lower plumbing fixture and fitting flow rates and, where municipal reclaimed water is available, to require that it be utilized in all buildings constructed within a jurisdiction. In addition to its mandatory water conservation and efficiency requirements, the IgCC also contains water related project electives which encourage the consideration and implementation of practices which further enhance water related performance beyond the minimum required by the code. Though most of the IgCC's requirements are mandatory, project electives are selected by the owner or designer from Table 303.1 (and Section 710 for water related electives) on a project by project basis.

REGIONAL WATER RESOURCE CHALLENGES

Geographic regions across the country have varying water resource issues along with unique regulatory constraints. Watershed boundaries, reservoirs, canals, water rights, and inter-state agreements all make up the regulatory and legal matrix. Local projects must work within the boundaries of state and local statutes. Many states have regional water regulatory agencies to plan, monitor, and manage a sustainable allocation of water resources. Increased frequency of droughts, population growth, increasing consumer water demand, agriculture, and industries which use large quantities of ground or surface water, hydraulic fracking, mining, widespread turf and landscape irrigation, and other factors are increasingly challenging the quality and quantity of available water. Many environmentalists feel that water conservation and efficiency is at least as critical a concern as is energy conservation and efficiency.

There is never a "one-size-fits-all" solution. Creative solutions may come about because of the unique conditions of a given site and an interpretation granted by the governing jurisdiction that meets the intent of the regulations. Successful projects do tend to have one thing in common: a quantifiable objective and a means of verifying outcomes.

SUSTAINABILITY AND MATERIAL RESOURCES

MATERIALS AND WASTE

Buildings consume 40% of the stone, gravel, and sand and 25% of the virgin wood in the world.[1] Construction and demolition wastes represent about 40% of the total solid waste stream in the United States.[2] Reducing the volume of construction waste can lower disposal costs and slow the filling of landfills, which consume environmental resources and can cause air and water pollution. Using materials with recycled content conserves raw materials and supports recycling of construction wastes so that they can be diverted from landfills. Many commonly used products are now available with recycled content, including metals, concrete, masonry, acoustic tile, carpet, ceramic tile, and insulation.

DESIGN WITH NATURE IN MIND

Vernacular Architecture and Natural Building Systems

For thousands of years, vernacular architecture around the world has been an integral part of local building traditions based on use of renewable resources and low impact technology. If we look at vernacular building traditions, there exist opportunities to learn of solutions to our building needs that draw less on the limited resources available to us. Vernacular architecture tends to tread gently on our fragile ecosystems, offering solutions that engender a profound connection between the builders, their environment, the materials used, and the wider community.

Vernacular buildings are built from local materials defined by the geology, ecology, and climate of the region. These structures are often highly practical, culturally appropriate, energy efficient, and blend with the landscape. These buildings carry many environmentally responsible attributes that are strived for in green buildings. Vernacular materials include wood, thatch, reed, bamboo, stone, and

[1]Roodman, David M. and Lessen, Nicholas, *Worldwatch Paper 124: A Building Revolution: How Ecology and Health Concerns are Transforming Construction* (Washington, DC: Worldwatch Institute 1995)

[2]LEED Reference Guide for Green Building Design and Construction (Washington, DC: USGBC, 2009)

Figure 7-1

Vernacular adobe traditions.

Figure 7-2

Wood is an example of a bio-based material.

Figure 7-3

Masonry is an example of a locally-produced material.

Figure 7-4

Corrugated metal is an example of recycled content material.

earth. Even today, an estimated one-third to one-half (2 to 3 billion people) of the world's population live or work in earthen structures (see Figure 7-1).[3]

The selection of building materials and products for a green building project can be a difficult process. The term *green building products* generally refers to materials or products that have low environmental impacts compared to conventional materials and products. Green building products can include any of the following:

- Biobased, natural, indigenous materials (see Figure 7-2)
- Local and regional materials (see Figure 7-3)
- Reused, recyclable, and recycled materials (see Figure 7-4)

[3]Dethier, Jean, *Down to Earth: Adobe Architecture* (New York: Facts on File, 1981) and Smith, Edward W. and Austin, George S., *Adobe, Press-Earth, and Rammed-Earth Industries in New Mexico*, (New Mexico Bureau of Mines and Mineral Resources, Bulletin 127, 1989)

Extraction, Processing, Manufacturing, and Transporting of Materials

The available type of construction materials and products have grown over time from relatively simple, locally available, natural, minimally processed resources to a combination of synthetic and largely engineered and manufactured products. Vernacular architecture evolved within the ecological limits of a given region with natural available resources such as earth, wood, rock, and a few low-technology products such as metals and glass. A minimum amount of energy and resources was required to extract, process, assemble, and transport the materials to the building site. Modern building materials have many unintended consequences associated

Figure 7-5

Life cycle of materials.

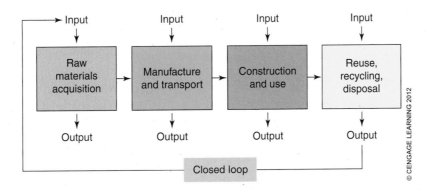

with extracting the raw materials, processing, and manufacturing, including ecological damage, diminishing scarce resources, emissions, and the energy required to operate the equipment for extraction, processing, manufacturing, and transporting the materials (see Figure 7-5).

Life-Cycle Assessment (LCA)

LCA is a methodology for assessing the environmental performance of a product over its full life cycle, often referred to as "cradle-to-grave" or "cradle-to-cradle" analysis. Whether it is explicit or implicit, every approach that considers materials from an environmental impact perspective uses some form of LCA. The LCA process is based on a life-cycle inventory, in which a researcher identifies and quantifies all of the raw materials and energy consumed in the extraction, manufacturing, packaging, transport, operation, cleaning, repair, maintenance, disposal, recycling, or disassembly and reuse of the product, as well as pollutants and by-products generated during the

Defining Green

Embodied Energy

The total energy consumed in the extraction, processing, manufacturing, transportation, and installation of a material or product. Products with greater embodied energy usually have a higher environmental impact due to the emissions and greenhouse gases associated with energy consumption. Recycled materials have lower embodied energy because they are not remanufactured from raw materials.

Figure 7-6

Embodied energy of selected building materials based on the manufacturing process.

Material	Embodied Energy	
	Btu/lb	MJ/kg
Aluminum	66,600	155
Portland cement	1,980	4.6
Concrete; typical in construction of buildings under three stories	408	0.95
With 25% fly ash as replacement for Portland cement	344	0.8
With 50% fly ash as replacement for Portland cement	279	0.65
Fiberglass insulation	12,000	28
Cellulose insulation	400–1,400	0.94–3.3
Steel; average of all steels	10,500	24

© CENGAGE LEARNING 2012

process. The energy consumed during this process is called is embodied energy. Each material has a distinct embodied energy profile (see Figure 7-6). Depending on the available data, this inventory may be comprehensive and detailed, or cursory, looking only to the most significant inputs and outputs.

Following the inventory, the LCA examines the environmental impacts of each of the material and energy flows. LCAs can assess materials and products, building systems and assemblies, or whole buildings. An LCA is as much an art as a science, because it involves the nearly impossible task of tracking ecological impacts at each step in the material production/use cycle.

The standard on which LCAs are based is from the International Organization for Standardization (ISO). The ISO 14000 standard provides a four-step LCA methodology: (1) goal and scope, (2) inventory analysis, (3) impact assessment, and (4) interpretation. LCA indicators often including the following:

- Fossil-fuel depletion
- Nonrenewable resource use
- Water use
- Global warming potential
- Stratospheric ozone depletion
- Ground-level ozone (smog) creation
- Nutrification/eutrophication of water bodies
- Habitat alteration
- Acidification and acid deposition (dry and wet)
- Toxic releases to air, water, and land

Defining Green

Life-Cycle Assessment (LCA)

An evaluation of the environmental impacts of a building system (e.g., use of resources and environmental consequences) throughout its life cycle, from raw material acquisition through manufacturing, construction, use, operation, recycling, and final disposal (end of life). The purpose is to identify opportunities to improve the environmental performance of buildings, materials, and components.

It is unquestionably a very valuable tool, but it is only as valuable and as perfect as its data and methodology permits it to be. LCAs can be used to evaluate any number of factors. All are valid if we understand their scope, intent and limitations. This applies whether we are discussing whole buildings, assemblies or specific material products. Even LCA standards that set criteria for the application of LCAs leave room for discrepancies. An LCA that focuses on a material's durability, but does not consider the impact on material extraction, may yield vastly different results than one that does the inverse. Thus it is critical that users understand the parameters and intent of the analysis so that materials are not selected or eliminated that may have unaccounted for benefits or impacts.

Closing the Material Loop

Since there is no waste in natural systems, it follows that the building materials cycle should be as waste-free as the laws of physics permit. Building material selection should follow the cardinal rules for an ideal closed-loop building materials strategy (see Table 7-1). The cardinal rules provide for the complete dismantling of the building and all of its components so that materials input at the time of the building's construction can be recovered and returned to productive use at the end of the building's useful life.

In addition to being deconstructable, requirements must also be in place to ensure that, when their service life is over, buildings will actually be deconstructed, and not simply be demolished. Products must not only be disassemblable, requirements and infrastructure must be in place to ensure that such products are actually disassembled and reused, recycled, or safely returned to the earth. Materials must not only be recyclable, requirements and infrastructure must be in place to ensure that such materials are actually recycled.

There are two general routes for recycling: technical and biological. The technical recycling route is associated with synthetic, manmade materials. These include metals, plastics, concrete, and non-wood composites. The biological recycling route is associated with natural materials. These include earth and biobased materials. This route is designed to allow nature to recycle building materials by turning them back into nutrients for ecosystems. Products of consumption are biological nutrients in the biological cycle (see Figure 7-7). Products of service are in the technical cycle (see Figure 7-8).

Material Selection and Reuse

When selecting building materials, as with water and energy resources, the first priority should be on reducing the quantity of materials needed for construction. Strategies to reduce material use include reduced building size, clustering and reduced building volume per square foot. LEED for Homes and ICC 700 incorporate provisions that incentivize smaller buildings. ICC 700 also awards points

Table 7-1
Cardinal Rules for a Closed-Loop Building Materials Strategy

1. Buildings must be deconstructable.
2. Products must be disassemblable.
3. Materials must be recyclable.
4. Products/materials must be harmless in production and in use.
5. Materials dissipated from recycling must be harmless.

© CENGAGE LEARNING 2012

Defining Green

Cradle-to-Cradle Design

As opposed to the conventional cradle-to-grave approach, cradle-to-cradle follows the process of nature in which waste equals food. The waste from one use or process becomes a resource for reuse or recycling into another use or process. The earth's major nutrients (carbon, hydrogen, oxygen, nitrogen) are cycled and recycled. This cradle-to-cradle biological system has nourished our thriving, diverse, and abundant planet for millions of years. As a result of industrialization, there are now two kinds of material flows on the planet:

1. Biological: Organic and recyclable as part of the biological cycle
2. Technical: Man-made materials useful for recycling and the industrial process

Manufactured materials and products can be composed of biological materials that biodegrade and become food for the biological cycle, or of technical materials that stay in a closed-loop cycle, where they recycle as valuable nutrients for industry.

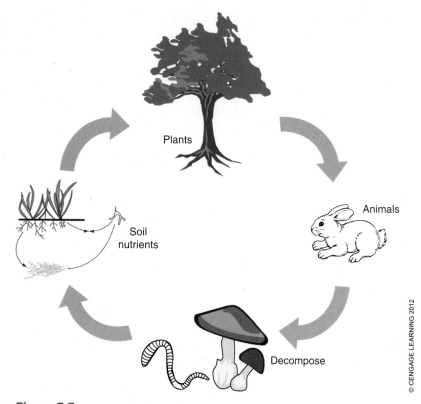

Figure 7-7

Biological nutrients come from products of consumption.

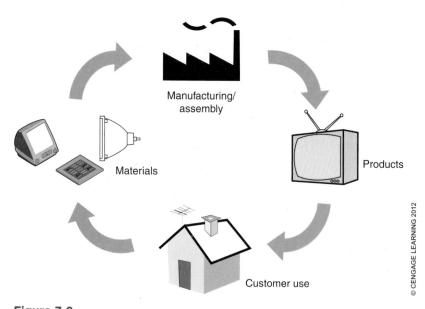

Figure 7-8

Technical nutrients come from building products of service.

for stacking stories, which reduces the amount of foundation and roof area in relation to total building floor area, and the structural members used for floors also serve as ceilings or roofs, thus reducing the amount of materials required. It is also important to base

© CENGAGE LEARNING 2012

building dimensions on material unit dimensions. This will help reduce waste from material cut-offs that is common in wood frame construction. Modular construction with its factory-engineered components reduces construction waste in a controlled environment.

A second priority should be to reuse materials and products from existing buildings. This involves the deconstruction of existing buildings. Deconstruction is the whole or partial dismantling of existing buildings for the purpose of recovering components for reuse. For some structural materials, however, testing may be required to verify their load bearing capacity. All materials do not perform indefinitely in the manner they were originally intended. Factors such as erosion, weathering and general use may take their toll. Reuse of materials for interior finish, decorative and other similar purposes may be less demanding where they are not subject to significant wear and tear.

A third priority should be to use products and materials that contain recycled content and that are themselves recyclable or to use products and materials made from local and renewable resources. All products, however, are not infinitely recyclable and are usually down-cycled to the point of diminished reusability. Take the time to understand the constituents of a material and how that might affect a product's ability to be recycled.

Material selection is about environmental material attributes and balancing those attributes with conventional material functions and uses. Moisture resistance, fire protection, toxicity and other factors must be taken into consideration. Table 601 of the International Building Code, Fire-Resistance Rating Requirements for Building Elements, classifies building construction types based on fire resistance of various elements. These provisions also account for durability (and sustainability) as an inherent element of the building code's classification of construction types. The green attributes of a material must always be considered in the actual context in which the material is to be used in a building and that materials ability to effectively perform its intended function in that context.

Building Reuse and Existing Buildings

By modifying an existing building and reusing as much of its structure and systems as possible, one can minimize the use of new materials, with their accompanying impacts of resource extraction, transportation, and processing energy, waste, and other effects. Of course, trade-offs must be made when considering a building for reuse. A leaky dilapidated existing building that is unsafe and resource inefficient may be a good candidate for replacement. Many buildings use many times their mass in the coal used in power generating plants that supply electricity. Depending on the condition, intended use and cost, such a building could be effectively retrofitted with safe, durable and energy efficient upgrades.

Defining Green

Deconstruction

The process of taking a building or structure, or portion thereof, apart, piece by piece, with the intent of recycling or salvaging as many of the materials, products, and components as possible.

Defining Green

Building Service Life

The intended service life or the period of time that a building or its component parts are expected to function without major repair.

Building Service Life

A building service life plan (BSLP) is a vital tool that supports the longevity of a building and its ability to stand the test of time. The BSLP typically addresses a period of not less than 60 years and include a maintenance, repair, and replacement schedule for each major building component. Values for component design service life and the maintenance, repair, and replacement schedule are often based on manufacturer data or other approved sources. The BSLP should account for future reconfiguration, dismounting, and disassembly of interior non-load-bearing walls, partitions, lighting and electric systems, suspended ceilings, raised floors, and interior air distribution systems for a minimum of 25 years.

Assessing a building design for building service life can be problematic. Although, there is service life data available for many common building materials such as concrete, steel or plumbing lines, there is limited data for other materials and components such as aged solar reflectance index (SRI) values for various roofing products. Service life is also often tied to set conditions of a manufacturer's warranty. Many warranties can be extended based on cost and the product track record. Besides mock lab tests, new products, materials and technologies haven't been around long enough to measure for durability.

Durability is a major component of building service life. Resistance to moisture, weather, wind, freeze-thaw and fire protection are attributes related to durability. As previously mentioned, many of these attributes are addressed in the building codes, though the codes do not encourage the use of one material over another based on their durability.

In many instances the useful life of a building will end far before its service life. Retail, restaurant and entertainment centers often demolish or execute major renovations to encompass new designs for marketability and services. For this reason, deconstruction and disassembly can be as important if not more important than building service life. Still, it is hard to argue against the fact that buildings should be designed for durability and long-term performance.

Construction Waste Management and Post-Occupancy Recycling

Recycling of construction and demolition debris reduces demand for virgin resources and in turn reduces the environmental impacts associated with resource extraction, processing, and transportation. By creating convenient recycling opportunities for building occupants, a significant portion of the solid waste stream can be diverted from landfills and incineration. Recycling of paper, metals, cardboard, and plastics reduces the need to extract virgin natural resources (see Figure 7-9). Recycling also reduces environmental impacts of waste in landfills and emissions from incineration. Landfills contaminate groundwater and encroach upon valuable green space. Land, water, and air pollution impacts can all be reduced by minimizing the volume of waste sent to landfills.

Figure 7-9

Post-occupancy collection and storage of recyclables.

COURTESY OF ANTHONY FLOYD

MATERIAL SELECTION TOOLS AND GUIDELINES

Material Attributes

There are many environmental attributes and claims about products containing recycled content, or being recyclable, biodegradable, or containing rapidly renewable content (see "Key Recycling Terms"). Product claims, however, are subject to being overstated and can be misleading. The U.S. Federal Trade Commission (FTC) regulates such claims through its "Green Guides" but has been significantly challenged with the ever-increasing green product market. A claim that a product is "recyclable" only means something if there is an accessible infrastructure for recycling it.[4] Professionals need to be familiar with the various certification programs in order to determine which attributes are important for the goals of their project, whether achieving a green building certification such as LEED or complying with the IgCC or ASHRAE 189.1. The bottom line is that the environmental beneficial attributes of a product may be outweighed by its negative attributes. All attributes must be considered, and may need to be weighted or evaluated in the context in which they are actually used in a building or system.

Defining Green

Construction Waste Management

Diversion of construction and demolition debris from disposal in landfills and incineration facilities. Redirection of recyclable and recoverable materials back to the manufacturing process and reusable materials to appropriate sites.

---

[4]Atlee, Jennifer, *Green Building Product Certifications* (Brattleboro, VT: BuildingGreen, Inc., 2011).

Biobased Material

A commercial or industrial material or product, other than food or feed, that is composed of, or derived from, in whole or in significant part, biological products or renewable domestic agricultural materials, including plant, animal, and marine materials, or forestry materials.

Sustainable Forestry

There are several sustainable forest programs in the North American market, including the Forest Stewardship Council (FSC), Sustainable Forest Initiative (SFI), American Tree Farm System (ATFS), and the Canadian Standards Association (CSA).

The Forest Stewardship Council (FSC), established in 1993 to standardize programs that had emerged during the 1980s, is an international nonprofit that manages an international standard for well-managed forests and a process for tracking and certifying products derived from those forests. The FSC addresses numerous aspects of sustainable forestry, including ecological functions, old-growth forests, plantations, restoration, native habitat, indigenous people's rights, and sound management for timber production (see Figure 7-10). The FSC has affiliate organizations in individual countries and different standards for different forest types and regions. Environmental groups generally view the FSC as the gold standard.

Key Recycling Terms

Recycled content: materials used in a product that have been recycled into the product. The benefit varies depending on the type of material, product, and recycling process. This sometimes requires a more comprehensive understanding of the life-cycle impacts of materials.

- **Pre-consumer (post-industrial) recycled content**: materials used in a product that are recycled from the manufacturing process.
- **Post-consumer recycled content**: materials used in a product that are recycled after consumer use.

Recyclable: a product's ability to be recycled after its useful life. Consideration should be given to the ease of deconstructing and recycling through an accessible infrastructure and cost-effective process.

Biobased materials: composed of biologically derived materials such as wood or agricultural by-products.

- **Rapidly renewable**: biobased materials having a harvest cycle of less than 10 years, such as bamboo or agricultural by-products including cotton and straw.

Biodegradable or compostable: a material that will break down into organic matter when exposed to certain conditions, reducing the load on landfills and becoming a beneficial supplement for soil. Examples include adobe, rammed earth, and straw bale construction.

Figure 7-10

Protection of native habitats in forestry.

COURTESY OF ANTHONY FLOYD

The Sustainable Forestry Initiative (SFI) was established in 1994 by the American Forest & Paper Association (AF&PA). The AF&PA is the primary wood products trade association in the United States and requires that all association members self-certify their compliance with its policies. The SFI has gradually distanced itself from the AF&PA and become a third-party certification program managed by an independent nonprofit. The SFI's current standard is more rigorous than previous versions and considers most of the issues addressed by its principal competitor, the FSC.

The American Tree Farm System (AFTS) was founded in 1941 by the American Forest Foundation and certifies forests as small as 10 acres (4 ha) for primarily nonindustrial landowners. The ATFS has long promoted responsible forestry but under fairly loose standards. In recent years it has moved toward more specific and prescriptive measures. Independent foresters accredited by the ATFS carry out certification. Unlike other certifications, the program doesn't have its own product label, but the SFI allows its logo on wood from ATFS-certified forests.

In 1993, the Canadian forest products industry turned to the Canadian Standards Association (CSA) to develop a standard for sustainable forest management. That effort culminated in 1996 with the release of Standard Z809, creating the CSA Sustainable Forest Management certification system. The CSA created this system as a process-based standard along the lines of ISO environmental management standards rather than a performance-based standard like the FSC and SFI, but it has since evolved to become similar to the SFI. The CSA is the dominant certification system in Canada, partly due to the large proportion of timberland in Canada that is publicly owned and government requirements to certify forest operations on public land. Like the SFI, the CSA has become more rigorous in recent years.

Green Product Certifications

The "UL" symbol of safety from Underwriters Laboratories and the Good Housekeeping Seal of Approval have influenced purchasing decisions for decades. The only way to ascertain the credibility of

Defining Green

Sustainable Forestry and Certified Woods

Woods from forests that manage resources to meet product needs while maintaining the long-term health of forest ecosystems, including wildlife habitats and biodiversity, and soil and water quality; minimizing the use of harmful chemicals; and conserving endangered and old-growth forests.

the green product is by certification. The success of major certification programs such as ENERGY STAR and interest from buyers in purchasing environmentally-friendly, responsible products have spurred a significant number of manufacturers to obtain green product certifications.

As compared to a single-green-attribute certification, multiple-green-attribute certifications assess a broad range of issues associated with a product, such as material composition, emissions, energy use, manufacturing impacts, and even the social responsibility of the manufacturer. Green product certifications help consumers identify products with the best environmentally responsible attributes (see Table 7-2).

Table 7-2
Green Product Certification Programs

Sustainable Forestry				
Forest Stewardship Council	Forest Stewardship Council (FSC)	Forest Products	Variety of labels for pure and percentage content	Third-party certification to regionally specific standards
Sustainable Forestry Initiative	Sustainable Forestry Initiative (SFI)	Forest products	Variety of labels for pure and percentage content	Third-party certification
American Tree Farm System	American Forest Foundation	Forest Products	Single standard for small U.S. landowners	Third-party certification
CSA Sustainable Forest Management System	Canadian Standards Association (CSA)	Forest Products	Single standard applied to specific forest areas	Third-party certification
Emissions Certifications				
California Section 01350	California Department of Health Services	Wide range of interiors products	n/a	Specification guidance on which other certifications are based
Greenguard	Greenguard Environmental Institute	Wide range of interiors products	Greenguard Indoor Air Quality, Greenguard for Children and Schools	Third-party certification
Floor Score	Scientific Certification Systems, Resilient Floor Coverings Institute	Non-textile flooring	FloorScore	Third-party certification
Indoor Advantage	Scientific Certification Systems (SCS)	Wide range of interiors products	Indoor Advantage, Indoor Advantage Gold	Third-party certification based on a variety of standards
Green Label	Carpet & Rug Institute (CRI)	Carpet, pad, adhesive	Green Label, Green Label Plus	Second-party certification
Energy				
ENERGY STAR	U.S. EPA and U.S Department of Energy	Range of products	n/a	Government label based on manufacturer data

Table 7-2 (*Continued*)

Multi-Attribute Standards and Certifications				
Sustainable Choice	Scientific Certification Systems (SCS)	Carpet; others expected	Silver, Gold, Platinum	Third-party certification, based on both consensus and proprietary standards
Cradle to Cradle (C2C)	McDonough Braungart Design Chemistry (MBDC)	Wide range of products	Biological, Technical Nutrients; Silver, Gold, Platinum	Second-party certification, based on a proprietary standard
SMaRT Consensus Sustainable Product Standards	Institute for Market Transformation to Sustainability (MTS)	Wide range of products	Sustainable, Silver, Gold, Platinum	Third-party certification
NSF-140 Sustainable Carpet Assessment	NSF International	Carpet	Bronze, Silver, Gold, Platinum	Standard, requiring third-party certification
Sustainable Furniture Standard	Business and Institutional Furniture Manufacturer's Association (BIFMA)	Furniture	Silver, Gold, Platinum	Standard, to which first, second, or third-party certification is possible
Green Seal	Green Seal	Wide range	n/a	Third-party certification
Ecologo/Environmental Choice	TerraChoice Environmental Marketing	Wide range of products	n/a	Third-party certification

COURTESY OF BUILDING GREEN (BUILDING GREEN.COM)

Pharos Project

The Pharos Project from the nonprofit Healthy Building Network is intended to provide comprehensive product analysis and ratings. Rather than developing a certification based on the best products in a given category, Pharos tries to define what an environmentally sustainable product should be—and compare existing products to that ideal. Pharos evaluates materials across 16 impact categories, such as energy and water use, air quality impact, and toxicity, as well as categories, such as occupational safety, social justice, and habitat impact, that are not typically emphasized (see Figure 7-11). Pharos presents this detailed information in an attractive graphic format intended to be easily readable. The Pharos Project, including the Pharos-Wiki, a forum for sharing product information, is designed to be user driven.

Sustainable Attributes Verification and Evaluation (SAVE) Program

The Sustainable Attributes Verification and Evaluation (SAVE) program provides building product manufacturers with verification that their products meet sustainability criteria as specified in green codes, standards, or rating systems, including the International Green Construction Code (IgCC), ASHRAE 189.1, and Green Globes. SAVE provides an independent evaluation that includes product testing at nationally recognized laboratories, inspection of the manufacturer's production process, and evaluation of data justifying attributes. The evaluation result in the issuance of a Verification of Attributes Report (VAR™).

Figure 7-11

Pharos impact categories.

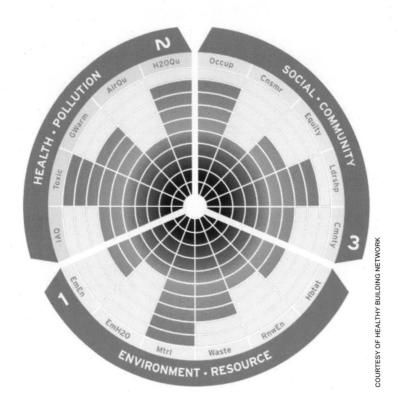

COURTESY OF HEALTHY BUILDING NETWORK

The ICC-ES has developed evaluation guidelines for use in evaluating manufacturers' claims about the environmental attributes of their products. These guidelines address the entire production stage, beginning with raw material acquisition and progressing through final manufacturing and packaging (cradle-to-gate). Evaluation guidelines include determinations for recycled content, biobased materials, solar reflective index (SRI), volatile organic compounds (VOCs), formaldehyde emissions, and certified woods (see Table 7-3).

Design professionals, code officials, and others can refer to the online SAVE Directory of Reports to find information on products that have been issued a VAR. Approved products are allowed to display the ICC-ES SAVE Mark (see Figure 7-12).

Figure 7-12

ICC-ES SAVE mark.

COURTESY OF THE INTERNATIONAL CODE COUNCIL

ATHENA Environmental Impact Estimator

The ATHENA Environmental Impact Estimator (EIE) is an LCA-based tool that focuses on the assessment of whole buildings or building assemblies such as walls, roofs, or floors. The EIE has a regional character, meaning that the user can select the project site from among 12 different North American regions. It accounts for material maintenance and replacement over an assumed building life, and distinguishes between owner occupied and rental facilities.

Building for Environmental and Economic Sustainability (BEES)

BEES is another prominent LCA tool for building materials and products that was developed by the National Institute of Science and Technology (NIST). BEES

Table 7-3

ICC-ES SAVE Evaluation Guidelines

EG101	Evaluation Guideline for Determination of Recycled Content of Materials
EG102	Evaluation Guideline for Determination of Biobased Material Content
EG103	Evaluation Guideline for Determination of Solar Reflectance, Thermal Emittance and Solar Reflective Index of Roof Covering Materials
EG104	Evaluation Guideline for Determination of Regionally Extracted, Harvested or Manufactured Materials or Products
EG105	Evaluation Guideline for Determination of Volatile Organic Compound (VOC) Content and Emissions of Adhesives and Sealants
EG106	Evaluation Guideline for Determination of Volatile Organic Compound (VOC) Content and Emissions of Paints and Coatings
EG107	Evaluation Guideline for Determination of Volatile Organic Compound (VOC) Content and Emissions of Floor Covering Products
EG107	Evaluation Guideline for Determination of Formaldehyde Emissions of Composite Wood and Engineered Wood Products
EG109	Evaluation Guideline for Determination of Certified Wood and Certified Wood Content in Products

allows side-by-side comparison of building products for the purpose of selecting cost-effective, environmentally preferable products, and includes both LCA and life-cycle costing (LCC) data. BEES provides data regarding air pollutants, indoor air quality, ecological toxicity, and human health for each material or product (see Figure 7-13). The overall score is divided into a separate environmental performance and economic performance score.

LEED-NC

LEED-NC is a rating system composed primarily of elective provisions that are assigned points. LEED-NC contains few mandatory provisions. Its mandatory provisions are its "prerequisites." The intent of LEED-NC's Materials and Resources section is to minimize the impact of building materials associated with the extraction, processing, transporting, and assembling of materials while reducing the waste generated during the course of construction. There is one prerequisite and eight optional credits (see Table 7-4).

Green Globes

Green Globes is a rating system containing elective provisions that are assigned point values. Green Globes does not contain mandatory requirements. The Green Globes Resources, Building Materials, and Solid Waste category assessment consists of online survey questions organized under the following six stages of the design construction process: (1) Pre-design, (2) Schematic Design, (3) Design

Figure 7-13

BEES environmental assessment categories.

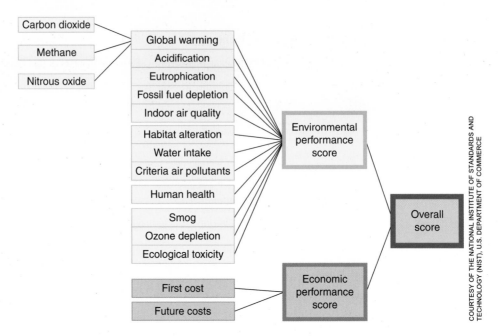

Table 7-4

LEED for New Construction: Materials and Resources

Credit	Prerequisite and Requirements
MRp1	Storage and Collection of Recyclables
MRc1.1	Building Reuse: Maintain Existing Walls, Floors and Roof
MRc1.2	Building Reuse: Maintain Existing Interior Nonstructural Elements
MRc2	Construction Waste Management
MRc3	Materials Reuse
MRc4	Recycled Content
MRc5	Regional Materials
MRc6	Rapidly Renewable Materials
MRc7	Certified Wood

Development, (4) Construction Documents, (5) Contracting/Construction, and (6) Commissioning. Each stage assesses material resource priorities with the objective of reducing environmental impacts (see Table 7-5).

LEED for Homes

LEED for Homes is a rating system composed primarily of elective provisions that are assigned points. LEED-NC contains few mandatory provisions. Its mandatory provisions are its "prerequisites." Similar to LEED-NC, the intent of LEED for Home's Materials and Resources section is to minimize the impact of building materials associated with the extraction, processing, transporting, and assembling of materials while reducing the waste generated during the course of construction. There are three prerequisites and six optional credits (see Table 7-6).

Table 7-5
Green Globes Resources/Materials Assessment by Project Stage

Pre-Design Stage

- Objective to minimize the environmental burden and embodied energy content of building materials and component assemblies
- Objective to optimize the use of resources
- Objective to minimize the waste from construction renovation, and demolition of the building
- Objective to minimize the waste generated during building occupancy

Schematic Design Stage

- Integration of systems and materials with low environmental impact
- Strategies to minimize the use of non-renewable resources
- Strategies to reuse parts of the existing building
- Design strategies for building durability, adaptability and disassembly
- Strategies to reuse and recycle demolition waste
- Facilities for recycling and composting

Design Development Stage

- Use of systems and materials with low environmental impact
- Strategies to minimize the use of non-renewable resources
- Strategies to re-use existing buildings
- Strategies for building durability, adaptability and disassembly
- Strategies to reuse and recycle demolition waste
- Facilities for recycling and composting

Construction Documents Stage

- Systems and materials with low environmental impact
- Materials that minimize consumption of resources
- Reuse of existing buildings
- Building durability, adaptability and disassembly
- Reuse and recycling of construction/demolition waste
- Facilities for recycling and composting

Contracting and Construction

- Optimization of the use of resources
- Minimization of waste from construction, renovation, and demolition

Commissioning

- Commissioning of mechanical systems to handle waste

COURTESY OF THE U.S. GREEN BUILDING INITATIVE (GBI)

CODES AND STANDARDS

GBI-01 Green Building Assessment Protocol for Commercial Buildings

GBI-01 is a rating system. It is also an ANSI standard. Chapter 10 of GBI-01 regulates resources/materials. GBI-01 requires that at least 29 percent of the 145 points available in the material/resource efficiency provisions in Chapter 10 be satisfied. (See Table 7-7)

Table 7-6

LEED for Homes: Materials and Resources

Credit	Requirements
MRp1.1	Framing Order Waste Factor Limit
MRc1.2	Detailed Framing Documents
MRc1.3	Detailed Cut List and Lumber Order
MRc1.4	Framing Efficiencies
MRc1.5	Off-site Fabrication
MRp2.1	FSC Certified Tropical Wood
MRc2.2	Environmentally Preferable Products
MRp3.1	Construction Waste Management Planning
MRc3.2	Construction Waste Reduction

Table 7-7

GBI-01 Green Building Assessment Protocol for Commercial Buildings

Chapter 10: Resources/Materials

Structural System and Envelope

Either performance or prescriptive methods may be used, not both.

1) Performance path: Max of 33 points awarded for LCA analysis as determined by application of The Green Globes LCA Credit Calculator for Building Assemblies

2) Prescriptive path: Max of 25 points available

    a) 1 to 8 points awarded where at least 1 to 20% of the materials in a building are recycled content materials

    b) 1 to 7 points awarded where at least 1 to 20% of the materials in a building are bio-based

    c) 1 to 5 points awarded where at least 1 to 20% of the materials in a building are in accordance with requirements for transportation of harvested, reclaimed salvaged or extracted materials

    d) 1 to 5 points where at least 1 to 20% of the materials in a building are in accordance with requirements for transportation of processed or manufactured materials

Furnishings, Finishes, and Fit-outs

- 1 point awarded, up to a max total of 4 points, for each ISO 14044 compliant LCA of a furnishing, finish or fit out
- 1 to 5 points awarded where at least 1 to 17% of furnishing, finish and fit-out materials are recycled content materials
- 1 to 4 points awarded where at least 1 to 16% of furnishing, finish and fit out materials are bio-based
- 1 to 6 points awarded where at least 10 to 60% of the materials in a project are wood-based

Reuse of Existing Structures

- 1 to 6 points awarded where at least 10 to 61% of building facades are reused, exclusive of windows and doors
- 1 to 6 points awarded where at least 10 to 95% of building structural systems are reused
- 1 to 6 points awarded where at least 10 to 95% of building non-structural elements, by area, are reused

Reduction, Re-use, and Recycling of Waste

- 2 to 6 points awarded where at least 25 to 75% of demolition and construction waste are diverted from landfills
- 1 point awarded where existing materials are reused for site development or landscaping
- 2 points awarded where features are incorporated to accommodate operational waste recycling programs
- 7 points awarded where a building service life plan is provided

Table 7-7 (*Continued*)

Resource Conservation through Design

- 7 points awarded where a building service life plan is provided
- 2 points awarded where the design professional documents how materials and raw materials were minimized in the building as compared to standard practices.
- 2 points awarded where the design professional documents how the building uses assemblies that perform multiple functions
- 3 points awarded where the building design facilitates demounting or disassembly of reusable materials.

Building Envelope

- 5 points awarded where roofing membrane assemblies and systems are designed and installed in accordance with manufacturer's requirements and are inspected in accordance with specified criteria
- 5 points awarded where all building flashing is detailed and installed in compliance with SMACNA's Architectural Sheet Metal Manual and field inspected in accordance with NIBS Guideline 3-06
- 5 points awarded where products for roof and wall openings are in accordance with the design pressure requirements of AAMA/WDMA/CSA 101/I.S.2/A440-05, flashings are detailed and installed in accordance with ASTM E2112-07, field inspections are conducted in accordance with NIBS Guideline 3-06, and water penetration field tests are conducted in accordance with ASTTM E1105-93
- 2 points awarded where vapour retarders are provided at all slab-on ground conditioned spaces
- 2 points awarded where dampproofing is applied to foundation walls, grade is sloped way from building at 5% for 10 ft min and a foundation drainage system is installed
- 4 points awarded where waterproofing membranes are installed at below grade slabs and foundation walls subject to hydrostatic pressure and they are inspected in accordance with NIBS Guideline 3-06 and ASTM D5957
- 5 points awarded where aluminum framed glazing systems, and EIFS, masonry vener and precast concrete exterior wall cladding systems and sealers are in accordance with 10.7.6.1
- 2 points awarded where rainscreen wall cladding is provided over frame wall systems in accordance with 10.7.7.1
- 3 points awarded where a continuous air barrier is provided in accordance with 10.8.1.1
- 3 points awarded where testing of continuous air barriers is conducted in accordance with 10.8.1.2
- 3 points awarded where vapor retarders are provided at the interior side of framed walls in accordance with the IECC 2007 Supplement or 2009 IBC in Climate Zones 5, 6, 7, 8 and Marine 4
- 3 points awarded where walls in unvented crawl spaces are insulated in accordance with 10.9.1.2

COURTESY OF THE GREEN BUILDING INITIATIVE (GBI)

ICC 700 National Green Building Standard

ICC 700 is a rating system. It is also an ANSI standard. Chapter 6 of ICC 700 addresses material resource efficiency. The only provisions that are mandatory in Chapter 6 of ICC 700 are those that are required in the International Residential Code or the International Building Code. The standard does, however, require increased points for material resource efficiency at each of its performance levels. A minimum of 45 points at the Bronze, 79 points at the Silver, 113 points at the Gold, and 146 points at the Emerald performance level must be acquired from the resource efficiency provisions of ICC Chapter 6. See Table 7-8 for a summary of these provisions.

Table 7-8
ICC 700 National Green Building Standard

Chapter 6: Resource Efficiency

Quality of Construction Materials and Waste

- Encourages the construction of smaller dwelling units by awarding 6 points to 2,500 s.f. dwelling units (the national average) up to 15 points for 1,000 square foot units, and penalizes larger units by requiring them to earn an additional 1 point for each 100 s.f. over 4,000 s.f. Multi-unit buildings must use a weighted average.
- 3 to 9 points awarded for structural systems or advanced framing techniques that are material efficient
- 3 points each awarded for floor, wall, roof, cladding or siding areas where 80% of the area is sized to reduce material cuts and waste. 1 point where 80% of penetrations or trim area is similarly sized.
- 4 points awarded where at least 2 stories are stacked and 2 points for each additional stacked story.
- 2 points each awarded where at least 50% of concrete, trim, wall covering systems or window skylight and door materials or assemblies are used that do not require site finishing. 5 points each are awarded where 90% of these materials or assemblies meet this criterion.
- 3 points awarded where material efficient frost-protected shallow, pier and pad or post foundations are utilized.
- 4 points awarded where at least 75% of exterior wall area is adobe, concrete, masonry, log or rammed earth.

Enhanced Durability and Reduced Maintenance

- 3 points awarded where the main entrance door system, and 1 point where other exterior doors, are protected by porches, roofs, awnings or recessing in accordance with its requirements, up to a total maximum of 5 points.
- 4 points awarded where roof overhangs are extended in accordance with Table 602.2, based on the average rainfall in the region.
- 4 points awarded where interior and exterior foundation drains are provided.
- 3 points awarded where a drip edge is provided at all roof edges.
- 4 points awarded where getters, downspouts or splash blocks and effective grading discharge water at least 5 feet from the building perimeter.
- 4 points awarded where a continuous physical foundation barrier or low toxicity treatment is provided in regions shown to have termite infestation potential by ICC 700 Figure 6(3).
- 2 to 6 points awarded where termite-resistant materials are used in accordance with its requirements.
- 4 points awarded where foundation waterproofing of the rubberized coating or drainage mat type is installed.
- 6 points awarded where flashing is installed at all the locations it lists.
- 3 points awarded where at least 90% of building roof surfaces are ENERGY STAR cool roof certified or landscaped.
- 3 points awarded where recycling collection areas are provided in each kitchen and a storage space for recycling containers is provided in a garage or outdoors.

Reused or Salvaged Materials

- 1 point awarded for each 200 s.f. of area of an existing building that is reused, up to a maximum of 12 points.
- 3 points awarded where salvaged materials and components are used for at least 1% of the project, based on cost.
- 4 points awarded where a sorting area is provided to facilitate the reuse of scrap building materials.

Recycled-Content Building Materials

- Awards points in accordance with ICC 700 Table 604.1 where recycled content materials are used for minor or major components.

Recycled Construction Waste

- 6 points awarded where a waste management plan to recycle or salvage at least 50% of construction and land-clearing waste is provided.
- 7 points awarded where at least 50% of construction waste is safely used on-site as fill or soil amendment.
- 605.3 awards 3 points where at least 2 types, and 1 point for each additional type, of material that is recycled from the list provided, up to a maximum of 6 points.

Table 7-8 (*Continued*)

Renewable Materials

- 3 to a maximum of 8 points awarded for the use of biobased materials other than wood.
- 3 to 4 points awarded where certified wood products are used.

Resource-Efficient Materials

- 3 to a maximum of 9 points awarded for the use of resource efficient materials such as thin brick, engineered wood, engineered steel and roof and floor trusses.

Indigenous Materials

- 2 points awarded for each indigenous material used for major elements in the building, up to a maximum of total of 10 points.

Life Cycle Analysis

- 3 points awarded for each LCA done to compare products or system, and 15 points for a whole building LCA, up to a total of 15 points.

Innovative Practices

- 1 point awarded for each 1% of materials whose manufacturers employ environmental management system concepts, and production facilities are ISO 14001 certified, up to a maximum of 10 points.

ASHRAE 189.1 Standard for the Design of High-Performance Green Buildings

The Atmosphere, Materials, and Resources section of ASHRAE 189.1 addresses the building's material-selection impact on the atmosphere and natural resources from the origin of the material components to the end of a product's useful life. The standard recognizes that these choices may contribute to pollution, habitat destruction, natural resource depletion, and unnecessary landfill diversions. The mandatory section must be met for all projects and an option for either a prescriptive or performance set of requirements must be followed to demonstrate compliance.

Mandatory Section

The mandatory requirements of Materials and Resources section address (1) the management of construction waste (see Figure 7-14), (2) the origin of building materials and products, (3) the selection of refrigerants, and (4) the storage and collection of recyclables and discarded goods. Waste diversion requires the tracking of construction and demolition waste generated on the project site and verifying that a minimum of 50% of that waste is either recycled or reused throughout the construction process. Tracking may be measured by either weight or volume and excludes hazardous materials, soil, and land-clearing debris. The waste can be stored and sorted either on the construction site or at a remote location.

Endangered wood species are prohibited from being used unless the trade of it conforms with the requirements of the Convention on International Trade in Endangered Species of Wild Fauna and Flora (CITES). Standard 189.1 prohibits CFC-based refrigerated in HVAC and refrigeration systems. To encourage the

Figure 7-14

Construction waste management.

COURTESY OF ANTHONY FLOYD

recycling of materials after building occupancy, a dedicated space must be designed into the building and reserved for the collection and storage of recyclable materials that are not hazardous. An additional area must be provided for the collection of fluorescent and HID lamps and ballasts complying with state and local hazardous waste requirements.

Prescriptive and Performance Options

The prescriptive path requires compliance with one of three categories: recycled content, regional materials, and biobased products. As an alternative to the prescriptive path, the performance compliance path option requires an LCA comparison between the proposed building and a baseline building. The proposed project building is required to show at least a 5% improvement in two of eight environmental impact categories.

International Green Construction Code (IgCC)

The IgCC is a code composed primarily of minimum mandatory requirements. As mentioned in previous chapters, the International Green Construction Code (IgCC) provides a comprehensive set of requirements intended to reduce the negative impact of buildings on the natural environment. Chapter 5 of the IgCC contains requirements related to building material selection, resource efficiency, and environmental performance (see Table 7-9).

Table 7-9

International Green Construction Code (Public Version 2)

Chapter 5: Material Resource Conservation and Efficiency

Section 502: Material And Waste Management

502.1 Construction material and waste management plan. Not less than 50% of non-hazardous construction waste shall be diverted from landfills. A construction material and waste management plan shall be developed and implemented to recycle or salvage construction materials and waste in accordance with this section.

502.2 Recycling areas for waste generated post certificate of occupancy. Waste recycling areas for use by building occupants shall be designed and constructed to accommodate recyclable materials based on the availability of recycling services.

502.3 Storage of lamps, batteries and electronics. Storage space shall be provided for discarded florescent lamps, HID lamps, batteries, electronics and other items requiring special disposal practices in the jurisdiction.

Section 503: Material Selection

503.1 Material selection and properties. Electrical, mechanical, plumbing, security and fire detection, and alarm equipment and controls, automatic fire sprinkler systems, elevators and conveying systems shall not be required to comply with Section 503.

503.2 Material selection. Not less than 55% of the total building materials (including modular furniture on tenant improvements) used in the project, based on mass or cost, shall comply with one or any combination of the following material properties:

- Used materials
- Recycled content materials (must contain at least 25% combined post-consumer and pre-consumer recovered material, and be recyclable or contain at least 50% combined post-consumer and pre-consumer recovered material),
- Recyclable materials (with a minimum recovery rate of 30%),
- Bio-based materials (with at least 50% bio-based content and labeled in accordance with SFI, FSC, PEFC, or equivalent fiber procurement system), or
- Indigenous materials (materials recovered, harvested, extracted and manufactured within 500 mile radius of the site, with special provisions for materials transported by water or rail).

503.1 Material selection and properties. Electrical, mechanical, plumbing, security and fire detection, and alarm equipment and controls, automatic fire sprinkler systems, elevators and conveying systems shall not be required to comply with Section 503.

Section 504: Lamps

504. Mercury content in lamps used in the buildings shall comply with the specified limits in this section.

Section 505: Service Life

505.1 Building service life plan. A building service life plan (BSLP) shall be included in the construction documents in accordance with this section. The building service life shall not be less than 60 years and shall be comprised of a dismantling, demounting, and re-use plan including interior adaptability.

Section 506: Moisture Control and Material Storage and Handling

506.1 Storage and handling of materials. Materials stored and handled on-site during the construction phases shall comply with applicable manufacturer's recommendations.

(Continues)

Table 7-9 (*Continued*)

506.2 Construction phase moisture control. Porous and fibrous materials and other materials subject to moisture damage shall be protected from moisture during the construction phase.

506.3 Moisture control preventative measures. Moisture preventative measures shall be inspected in accordance with Sections 902 and 903 for the following categories: 1) Foundation sub-soil drainage system; 2) Foundation damp-proofing and water-proofing; 3) Flashings: Windows, exterior doors, *skylights*, wall flashing and drainage systems; 4) Exterior wall coverings; and 5) Roof coverings, roof drainage, and flashings

Section 507: Strawbale Construction

507.1 Scope. Where applicable, this section shall govern the use of baled straw as a building material.

COURTESY OF THE INTERNATIONAL CODE COUNCIL

All of the provisions if the IgCC discussed in Table 7-9 are mandatory except where they are indicated to be project electives. Project electives are selected from IgCC Table 303.1 by the owner or designer and may vary from project to project. The total number of project electives required on each project is determined by the jurisdiction in Table 302.1.

THE 3 RS: REDUCE, REUSE, RECYCLE

The selection of building materials has profound consequences related to environmental impacts and energy consumption. Green buildings mitigate these impacts through resource-efficient design and construction. Resource efficiency boils down to the environmental mantra of the 3 Rs: reduce, reuse, and recycle. Do more with less by designing and constructing buildings that minimize waste and unnecessary materials. Reuse materials by selecting materials with recycled content and salvaged components where possible. Select biobased (wood and agricultural) products that are certified from sustainably managed forests. Design for an extended building service life by considering the durability of major building components, including the mechanical, plumbing, and electrical systems. Account for the reuse of buildings or their components by considering adaptability, deconstructability, and recyclability.

It is also important to recognize that different regions of the country have different environmental challenges while offering unique opportunities based on regional material availability and construction practices. Projects should always seek first to select materials that are extracted or harvested, manufactured, and assembled within the region. This will reduce environmental impacts associated with transportation, shorten the material supply chain for better environmental accountability, and strengthen the local economy.

ENERGY EFFICIENCY AND ATMOSPHERIC QUALITY

ENERGY CONSUMPTION AND EFFICIENCY

Energy consumption is the single most important green building issue. Buildings consume about 40% of primary energy and 73% of the electricity produced in the United States annually.[1] Energy-efficiency measures are intended to reduce the use of and dependence on nonrenewable energy resources, such as fossil fuels and nuclear energy. Conventional fossil-based electricity generation emits CO_2, which is a key factor in global climate change. Coal-fired power plants emit one-third of the country's anthropogenic nitrogen oxide (key factor in smog) and two-thirds of the sulfur dioxide (key factor in acid rain).[2] They also emit more fine particulate matter than any other activity in the United States, which contributes to cancer and respiratory-illness-related deaths annually. Natural gas is a major source of nitrogen oxide and greenhouse gas (GHG) emissions.[3] Nuclear power plants increase the potential for catastrophic accidents and raise significant waste transportation and disposal issues. There are also ecological impacts associated with coal and uranium mining.

Energy-efficiency measures reduce GHGs, improve regional and global air quality, and mitigate the potential effects of global warming. These measures also mitigate the need for power plant construction and expansion. As part of energy efficiency and reducing reliance on fossil fuels, there is a move toward net-zero-energy buildings (buildings that use on-site renewable energy to produce at least as much energy as they consume) and buildings that utilize only net-zero grid energy (wherein the utility grid provides 100% of power from renewable energy sources).

The International Energy Conservation Code (IECC), as its very name indicates, is a conservation code, and the requirements serve as baseline requirements for many green and sustainable rating systems that simply ramp up the baseline provisions of energy codes, often by specific percentages. Although energy code baseline requirements have been significantly improved in recent editions, they do not encourage the use of renewable power, or zero-net-energy buildings that are as effective as those in green building rating systems and hybrid codes like the IgCC.

The U.S. Department of Energy (DOE) has indicated that the scale of the energy challenge is massive and its implications are wide ranging. In this regard,

---

[1] 2008 Buildings Energy Data Book. 2008. http://buildingsdatabook,eren.doe.gov/ChapterIntro6.aspx (accessed September 2011).

[2] U.S. Environmental Protection Agency. "Clean Energy: Air Emissions." http://www.epa.gov/cleanenergy/ energy-and-you/affect/air-emissions.html (accessed September 2011).

[3] Ibid.

voluntary programs are relatively ineffective in the mass market. Higher efficiency baseline energy requirements are our only truly effective approach for addressing these concerns. Energy codes become mandatory when they are adopted, whereas green building rating systems, as well as green codes and standards that are adopted on a voluntary rather than mandatory basis are not.

DESIGN WITH NATURE

Vernacular Architecture and Natural Energy Systems

Ralph Knowles, professor emeritus at the University of Southern California, describes the methods people use to adapt to nature: migration, which can mean movement across many miles or just within a room of house; transformation, such as by simply opening a window to air out a room or opening a curtain to let light in; and metabolism, or producing energy, such as by lighting a fire in a fireplace or adjusting a thermostat.[4] Historical examples include the Bedouins tents of North Africa, the toldos (translucent canvas covers) courtyards of houses in southern Spain, and the shoji screens (sliding doors) in traditional Japanese houses (see Figure 8-1).

The properties of matter and energy must be considered in order to fully understand climatic phenomena. Heat, radiation, pressure, humidity, and wind interact to create macro- and micro-climatic conditions on the earth's surface. Before the advent of industrialization and mass production, humans depended on natural sources of energy and available local materials in building habitats.

For thousands of years, people have learned to interact with the local climate and ecosystem. The interaction of climate, seasons, and local resources shapes

Figure 8-1

Shoji screens in a traditional Japanese house.

© ISTOCKPHOTO/DAVID KERKHOFF

---

[4]Knowles, Ralph L., *Ritual House: Drawing on Nature's Rhythms for Architecture and Urban Design* (Washington, DC: Island Press, 2006).

the rhythm of their lives as well as the design of habitats. Notice how the gable roof pitch decreases as the rate of precipitation decreases.[5] In Northern Europe and areas subject to heavy rain and snow, gable roofs are steep, whereas in the summer lands of the south, the pitch deceases. In the hot climates of the North African coast, the roofs become quite flat, often serving as a place to sleep during summer nights, when the effects of night sky radiation offer a cool respite from the higher temperatures of the ground surface. Still further south, in the tropical rainfall zones, the roofs are again steep to provide protection from the torrential downpours typical of the region (see Figure 8-2).

Traditional architecture grew out of countless experiments and the experience of generations of builders who continued to use what worked and rejected what didn't. They were passed on in the form of tradition and rules for selecting sites, orienting the building, choosing the right materials, and locating window size and locations. The synergistic effects of these elements are essential. If one building element changes, however slightly, it could diminish the building's capability to provide comfort and a satisfactory solution to the local climatic conditions.

The importance of climate is clear. All living organisms depend on climate for their existence and adapt themselves to these environmental conditions.[6] Plants that live in the tropics cannot live in the arctic, nor can arctic plants live in the tropics, unless located in a microclimate such at the top of a high equatorial mountain. Most organisms are limited to climatic or seasonal conditions of a narrow range. However, not all species are so limited. Many animals can regulate their own internal body temperatures and maintain it even during fluctations of the air temperature. Humans have an elaborate and sensitive biological mechanism

Figure 8-2

Steeply pitched thatched roof.

© ISTOCKPHOTO/CHRIS CRAFTER

---

[5]Fathy, Hassan, *Natural Energy and Vernacular Architecture* (Chicago: The University of Chicago Press, 1986).

[6]Ibid.

Figure 8-3

Comfort and Thermal Exchange: Our bodies have a number of biological mechanisms to regulate heat flow to ensure that heat loss equals heat generated and that thermal equilibrium is achieved at 98.6°F. Any deviations can cause severe stress and possible death.

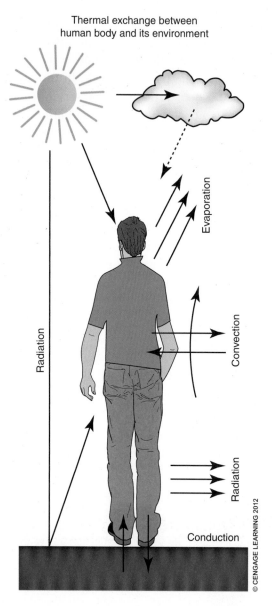

Thermal exchange between human body and its environment

Evaporation

Convection

Radiation

Radiation

Conduction

© CENGAGE LEARNING 2012

involving the secretion of sweat and the distribution of blood that keeps them at about 98.6°F (see Figure 8-3). Only 10 to 15 degrees higher or 20 degrees lower can cause death. In general, warm-blooded animals can survive wider variations than cold-blooded ones and some species, including humans, manipulate their environment to produce a favorable microclimate.

A principal purpose of building is to create a microclimate for human comfort. The goal throughout human history has been to produce an environment favorable to comfort, security, and, in more recent times, productivity. Early humans built shelters for protection from the uncertain dangers of nature while recognizing the subtleties of nature. The application of modern technology in the design and construction of buildings without regard to the subtleties of nature often results in an overuse and waste of increasingly limited resources, and is in the long run a detriment to society. A basic understanding of our bioclimatic environment is essential for environmentally appropriate and efficient building solutions.

BASIC ENERGY PRINCIPLES

Heating, cooling, and lighting a building involves the addition and subtraction of heat. Heat is a form of energy that demands a basic understanding in the design, operation, and performance of buildings. Its unit of measure in the United States is the British thermal unit (BTU). A BTU is defined as the amount of heat needed to raise one pound of water by one degree Fahrenheit.

Heat flows in one direction, from hot to less hot (see Figure 8-4). Without a temperature differential, there can be no transfer of energy. Heat energy can be sensible or latent. Sensible heat is dry heat and is the result of molecular vibration in a substance. As molecular vibration speeds up, the measured temperature rises, meaning hotter objects have a greater intensity of molecular vibration. Latent heat is heat released or absorbed by a substance during a change of phase such as a material changing state from a solid to a liquid, liquid to gas, or vice versa. Evaporation and condensation of liquid, be it water or some other form of refrigerant, make it possible to move heat from one location to another. The release or absorption of latent energy during change of phase is what makes steam heat and the refrigeration process possible. Removal of latent heat through humidity control is central to air conditioning.

Heat transfer occurs by conduction, convection, and radiation (see Figure 8-5). If two objects with varied temperatures are in direct contact with each other, heat will be transferred by conduction from the warmer to the cooler temperature. A measure of the rate at which heat is transferred through a material by conduction

Defining Green

Heat Transfer

Heating and cooling a building involves the addition and subtraction of heat. There are three primary mechanisms of heat transfer: conduction, convection, and radiation. Conduction requires direct physical contact. Convection requires a movement of a fluid such as air or water. Radiation occurs between surfaces from higher to lower temperatures.

Figure 8-4

Heat Flow: Although all objects absorb and emit heat, there will be a net flow of heat from warmer to cooler objects.

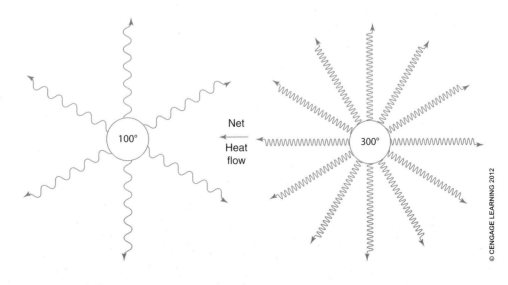

© CENGAGE LEARNING 2012

Figure 8-5

Heat Transfer: Heat is transferred by conduction, convection, and radiation.

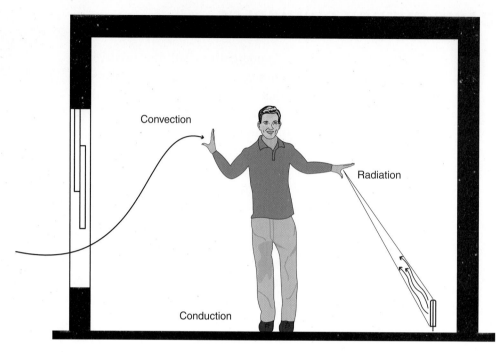

is known as the U-factor. A lower U-factor indicates a lower rate of heat transfer, which is the goal of building insulation in limiting heat transfer between inside and outside. The insulating value of a material is usually indicated as a measure of its ability to resist heat transfer and is expressed as R-value, which is the inverse of the U-factor. The higher the R-value, the greater the material's resistance to heat flow. Convection is the second form of heat transfer. Liquids and gases expand when they gain heat and contract when they lose heat. Upon expansion, a substance becomes lighter and less affected by gravitational forces. As such, warmer air naturally rises, creating convective currents that transfer heat. Cool, denser air naturally falls to the ground (see Figure 8-6). For building applications, natural convection moves heat from lower to higher locations within an enclosed space. As a consequence, lower locations are cooler. Forced convection occurs when air is moved by wind or mechanically by a fan or when liquid is moved by a pump.

The third form of heat transfer is radiation. All objects (above absolute zero) emit and absorb electromagnetic energy by means of radiation. The frequency of the wavelength at which an object radiates is directly related to that object's temperature. The hotter the object, the more heat will be emitted. The sun, the ultimate source of heat, emits radiation in various wavelengths of the electromagnetic spectrum. Radiant energy is not affected by gravity. It travels in all directions from its source (see Figure 8-7). Short wave radiation can be transmitted through such materials as glass. Radiant energy that is not transmitted is either absorbed by the material or reflected away. The interaction between radiant energy and a material is affected by the properties of the material, its surface, and the wavelength (short or long) of the radiant energy.

Materials continually exchange radiant energy between warmer and cooler objects. Warmer objects in a room will give off heat to cooler objects. This phenomenon is called mean radiant temperature (MRT), which is the rate of exchange of radiant

Figure 8-6

Convection Currents: Natural-convection currents result from warmer lighter air rising and cooler dense air falling.

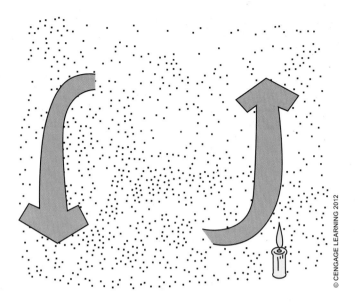

Figure 8-7

Radiant Energy: Since radiant energy is not affected by gravity, it travels in all directions from it's source.

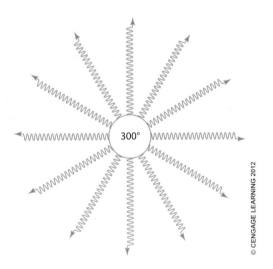

energy between two objects based on the temperature difference between the objects and their angular relationship (see Figure 8-8).

Comfort factors for building occupants include air temperature, humidity, air movement, mean radiant temperature, the presence or lack of solar radiation, and personal factors of clothing, weight, metabolic rate, and level of physical activity. Comfort is assumed for 80% to 90% of the building occupants when the temperature is between 68°F (20°C) and 82°F (27.8°C) and relative humidity is between 20% and 50%. As humidity rises from 50% to 80%, the acceptable upper temperature range declines from 82°F (27.8°C) to 75°F (23.9°C) (see Figure 8-9). Air movement from a ceiling fan or natural ventilation can increase the upper limit of the comfort zone by nearly 10°F when relative humidity is between 20% and 50%.[7] There are also psychological factors that affect comfort level. Degree of control over one's personal environment has been shown to increase occupant satisfaction under certain

---

[7]Keeler, Marian and Burke, Bill, *Fundamentals of Integrated Design for Sustainable Building* (Hoboken, NJ: Wiley, 2009).

Figure 8-8

Mean Radiant Temperature (MRT): On a winter day we receive radiation from hot surfaces and transmit radiation to cool surfaces. The MRT at any point is a result of the combined effect of a surface's temperature and angle of exposure.

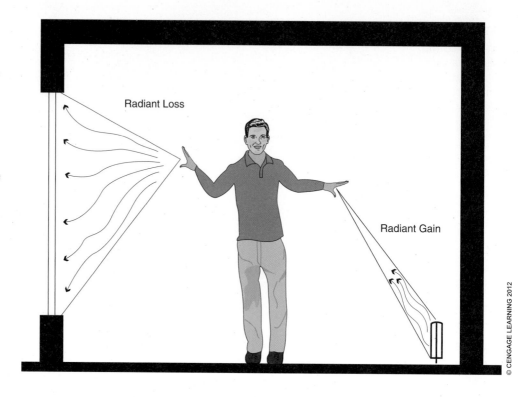

Radiant Loss

Radiant Gain

© CENGAGE LEARNING 2012

Figure 8-9

Human Comfort Zone: Shading from the sun, air movement, humidification, and the use of thermal mass can extend the limits of the human comfort zone.

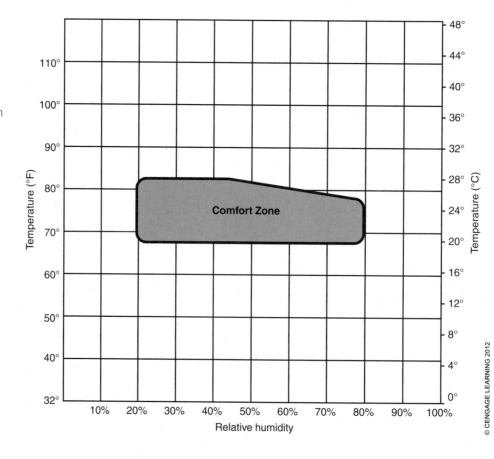

Comfort Zone

© CENGAGE LEARNING 2012

Figure 8-10

Climate zone map.

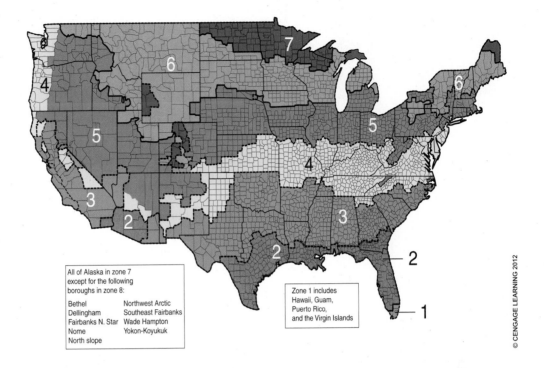

All of Alaska in zone 7
except for the following
boroughs in zone 8:

Bethel	Northwest Arctic
Dellingham	Southeast Fairbanks
Fairbanks N. Star	Wade Hampton
Nome	Yokon-Koyukuk
North slope	

Zone 1 includes
Hawaii, Guam,
Puerto Rico,
and the Virgin Islands

© CENGAGE LEARNING 2012

thermal conditions. Light, color, and sound can affect comfort. A fountain suggests coolness in the color, texture, and sound of water. Color creates a mood. Red vitalizes. Green is cooling, soothing, and calming. The color blue is cold.

Climate affects human comfort, given the way we occupy buildings and the activities we conduct inside those buildings. Characteristics of climate include data on temperature, relative humidity, wind speed and direction, sky conditions, and precipitation. As mentioned in Chapter 2, there are many different climate zones in North America. The U.S. DOE divides North America into five broad categories: (1) Cold/Very Cold, (2) Hot/Humid, (3) Hot Dry/Mixed-Dry, (4) Mixed-Humid, and (5) Marine. For the purposes of energy efficiency, ASHRAE divides the United States into eight climate zones (see Figure 8-10). Energy code requirements, such as the iECC for new buildings, are based on these zones.

PASSIVE DESIGN

Passive design is the measure of the building's heating, cooling, lighting, and ventilation systems, that rely on sunlight, wind, vegetation, and other naturally occurring resources on the building site. A building that has been well designed in a passive manner could be disconnected from its off-site energy sources and still be reasonably comfortable and functional. An optimized passive design can significantly reduce energy costs, reliance on non-renewable fuels, and the environmental impacts associated with heating, cooling, ventilation, and lighting of buildings.

Passive design strategies are based on local climate and site conditions. It involves building orientation, building aspect ratio, building mass, skin- versus internal-heat dominated loads, daylighting, and ventilation. In most cases, the ideal orientation of a building on a site is to locate the long side on an east-west axis in

Defining Green

Insulating Effect of Thermal Mass

Thermal mass can store heat for passive heating and can be used as a heat sink for passive cooling, thereby eliminating peak demand for air conditioning. The time-lag property of mass materials can be used to reduce both the peak load and the total internal heat gain during the summer. Indigenous buildings in hot and dry climates were usually built of stone, soil, or adobe to reduce heat gain by delaying the entry of heat into the building until the sun has set. For optimal application: (1) never use thermal mass without insulation; (2) mass is best used on the inside of a building; and (3) do not insulate the mass from the indoors.

order to minimize solar heat gain on the east and west walls during the summer. The building aspect ratio is the ratio of a building's length to its width, which is an indicator of the shape of a building. A building in the colder climate (northern United States) should have an aspect ration of close to 1.0, meaning it should be square in shape. For buildings in a warmer climate of the lower southern latitudes, the aspect ratio increases, meaning the building becomes longer and narrower. Windows on the east and west walls are minimized to minimize summer heat gain in the mornings and afternoons. South-facing walls can be easily shaded with roof overhangs for shading purposes.

Thermal mass is important for both passive heating during the winter and passive cooling during the summer. Mass materials include brick, concrete, masonry, and earthen materials such as adobe and rammed earth (see Figure 8-11). These materials have a high capacity to hold heat. For passive heating, the geometry of the building is arranged to allow interior mass materials in the floor and walls to be struck by the sun, thereby absorbing and storing heat. This heat is released during the evening and into the night. For passive cooling, interior mass floors and walls are protected from sun exposure and absorb heat from the interior space, thereby making one feel cooler. Although mass can keep a building cool during the day, the heat absorbed by the mass must be released during the evening by either (1) opening windows for nighttime cross-ventilation in climates with cooler nighttime temperatures or (2) air conditioning to remove the heat from the walls and floors. The thermal lag in mass essentially shifts the inside peak temperature to early and late evening. The ideal passive design strikes a balance between passive heating in the winter and passive cooling in the summer. This requires careful consideration of orientation, fenestration (doors and windows), massing, and shading.

From a heat gain perspective, buildings can be divided into two types: skin-load dominated and internal-load dominated (see Figures 8-12a and b). The skin-load dominated building is much more sensitive to climatic conditions due to a large skin-to-volume ratio and low internal heat sources. This building type includes homes and small commercial structures. The internal-load-dominated building is less sensitive to climatic conditions due to a small surface-to-volume ratio and high internal heat gains from machines, equipment, lights, and people. This building type includes medium to large commercial buildings such as offices, malls, hotels, hospitals, and factories. Knowing whether a building is skin-load dominated or internal-load dominated is vital in determining appropriate design strategies that affect the building envelope and heating/cooling requirements based on climate zone and time of year.

Daylighting is often one of the simplest and most effective methods to reduce the use of electricity. The greatest annual electrical demand usually occurs during the sunny summer afternoons, when the air conditioning is working at full capacity. Although the sun creates this peak cooling load, it simultaneously provides a maximum amount of daylight. Often most of the electric lights, which consume about 40% of the total building energy, are not needed. The maximum demand

Figure 8-11

Rammed earth construction.

COURTESY OF ANTHONY FLOYD

COURTESY OF ANTHONY FLOYD

Skin-load

COURTESY OF ANTHONY FLOYD

Internal-load

Figure 8-12

Skin-load dominated buildings are more sensitive to climate because they have a large surface-area-to-volume ratio and modest internal heat sources such as people, lights, and equipment. Examples include houses, small office buildings, and this small physical therapy facility. Internal-load dominated buildings are less sensitive to climate because they have a small surface-area-to-volume ratio and large internal heat gains from people, lights, and equipment. Examples include medium-to-large commercial buildings including hotels, hospitals, and factories.

for electrical power can therefore be reduced up to 40% by the optimization of daylighting.[8] Daylighting requires the careful selection and placement of fenestration for the proper distribution and quality of daylight within an optimal 15-foot perimeter (see Figures 8-13a, b, and c). The general goal of daylighting is the same as for electric lighting, which is to supply sufficient light of high quality while minimizing direct glare, veiling reflections, and excessive brightness ratios. Both the orientation and form of a building are critical to a successful daylight scheme.

---

[8]Lechner, Norbert, *Heating, Cooling, Lighting: Sustainable Design Methods for Architects* (Hoboken, NJ: Wiley, 2009).

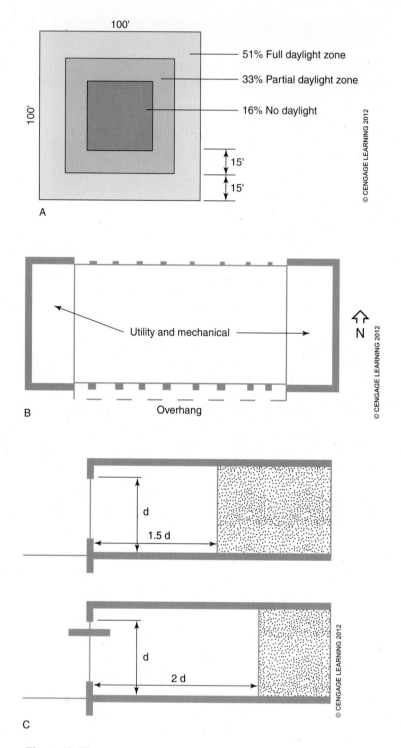

Figure 8-13

(A) Daylight Zones: In multi-story buildings, a 15-foot perimeter zone can be fully daylit and the next 15 feet zone that can be partially daylit. By adding a courtyard or atrium in the center, the inner 15 foot zone can be zfully daylit. (B) Building Orientation: The ideal orientation for daylighting is with all windows facing south and north. The south gets consistent sunlight throughout the day and year. The north gets a constancy of high quality light with very little glare. East and west facing windows are the least desirable due to excessive glare and available sunlight for only half of each day. (C) A rule of thumb for daylight penetration is 1.5 times the height of a window and 2 times the height of a window with a light shelf for south-facing windows under direct sunlight.

Not only must the external building form be considered, but also the shape of internal spaces. Internal partition height, glass window walls, ceiling heights, depth of floor plate, and room finishes will all contribute to the effectiveness of daylight.[9] See Chapter 9 for additional discussion on daylighting.

BUILDING SYSTEMS FOR ENERGY EFFICIENCY

Building Envelope

After passive design is used to optimize climate, site conditions, building orientation, and shape, heat transmission must be minimized through an air-tight and thermally resistant envelope. The building envelope must control solar heat gain, conduction, and infiltration. The major components of the building envelope consist of opaque walls, fenestration (doors and windows), floors, and roofs.

Wall and ceiling assemblies are usually the dominant component of the building envelope. Insulation limits the transfer of heat across wall and ceiling assemblies, below raised floors, slabs on grade, basements, and foundation walls. The minimum R-value and maximum U-factor are set by energy codes and standards such as the IECC or ASHRAE 90.1 Energy Standard for Buildings Except Low-Rise Residential Buildings.

The recommended R-value and U-factor is a function of the number of heating degree days (HDDs) and cooling degree days (CDDs) in a given climate. Both HDD and CDD are measures of how much energy is required for heating or cooling based on the climate zone. An important consideration is the placement of insulation and depending on climate zone, the strategic use of thermal mass in the wall assembly (see Figure 8-14). Insulation performs best when placed closest to the exterior wall surface. Thermal mass such as concrete, masonry, and earthen materials provide the greatest benefit when placed closest to or exposed to the interior conditioned space.[10] Insulation minimizes thermal transmission from outside ambient

Defining Green

Skin-Load Dominated Versus Internal-Load Dominated Buildings

The skin-load dominated building is greatly affected by climate because it has a large surface-area-to-volume ratio and small internal heat sources. These buildings include residences and small commercial buildings. On the other hand, the internal-load dominated building is generally less affected by climate because it has a small surface-area-to-volume ratio and large internal heat gains from equipment, lights, and people. These buildings include large commercial and school buildings, auditoriums, theaters, and factories.

---

[9]Lechner, Norbert, *Heating, Cooling, Lighting: Sustainable Design Methods for Architects* (Hoboken, NJ: Wiley, 2009).

[10]Ibid.

Figure 8-14

Wall Assembly with Mass: Wall assembly in hot-dry climate with mass exposed to the interior conditioned space. Insulation is placed on the exterior.

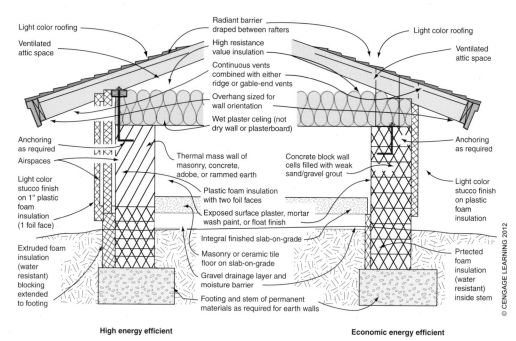

Light color roofing

Ventilated attic space

Radiant barrier draped between rafters

High resistance value insulation

Continuous vents combined with either ridge or gable-end vents

Overhang sized for wall orientation

Wet plaster celing (not dry wall or plasterboard)

Light color roofing

Ventilated attic space

Anchoring as required

Airspaces

Light color stucco finish on 1" plastic foam insulation (1 foil face)

Anchoring as required

Thermal mass wall of masonry, concrete, adobe, or rammed earth

Concrete block wall cells filled with weak sand/gravel grout

Plastic foam insulation with two foil faces

Exposed surface plaster, mortar wash paint, or float finish

Light color stucco finish on plastic foam insulation

Extruded foam insulation (water resistant) blocking extended to footing

Integral finished slab-on-grade

Masonry or ceramic tile floor on slab-on-grade

Gravel drainage layer and moisture barrier

Footing and stem of permanent materials as required for earth walls

Prtected foam insulation (water resistant) inside stem

© CENGAGE LEARNING 2012

High energy efficient   **Economic energy efficient**

Figure 8-15

Window Performance: A low solar heat gain coefficient (SHGC) value generally means a spectrally selective low-e coating protects the window from infrared radiation (heat transmittance).

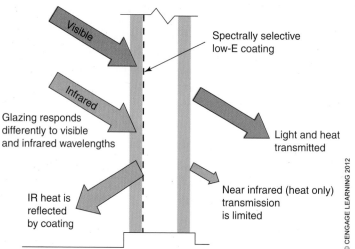

Visible

Infrared

Spectrally selective low-E coating

Glazing responds differently to visible and infrared wavelengths

Light and heat transmitted

IR heat is reflected by coating

Near infrared (heat only) transmission is limited

© CENGAGE LEARNING 2012

conditions, and thermal mass helps to balance the indoor temperatures based on the principle of mean radiant temperature (MRT).

Besides providing natural light, views, and ventilation (when operable), windows serve as a thermal barrier to less favorable outside conditions. Window performance is measured by solar heat gain coefficient (SHGC), U-factor, visible transmittance (VT), and infiltration of the window assembly. SHGC is the fraction of solar heat that enters the window and becomes heat (see Figure 8-15). It is expressed with a value between 0 and 1. The lower the SHGC, the less solar heat the window transmits through the glazing from the exterior to the interior. U-factor, as previously discussed, is a measure of heat transfer for a specific material or assembly. It is usually expressed with a value between 0 and 1 with the lower value representing minimum heat transfer. VT refers to the percentage of the visible spectrum of sunlight that is transmitted through the glazing. It is expressed with a value between 0 and 1, with the higher value representing the maximum level of

Figure 8-16

National Fenestration
Rating Council
(NFRC) label.

COURTESY OF THE NATIONAL FENESTRATION RATING COUNCIL

visible light. A clear, single-pane window has a VT of 0.90, meaning that it admits 90% of the visible light. A low VT value may be appropriate for an office building or where reduced interior glare is needed.

Because windows are not as thermally resistant as walls, increasing the window area to maximize daylighting reduces the thermal performance of the overall building envelope. In balancing the percentage of windows, one must consider the building orientation, shading, function of interior spaces, and, most important, the energy performance characteristics of the windows in the design process. The National Fenestration Ratings Council (NFRC) provides a national rating system for measuring the energy performance of fenestration products, including windows, doors, and skylights. NFRC certification provides a product label for U-factor, SHGC, VT, and air leakage (see Figure 8-16).

A tight building envelope will reduce uncontrolled movement of air between interior and exterior. This increases the effectiveness of passive-design strategies and reduces the energy needed to supplement passive design. An air barrier is essential for a tight building envelope. Any material that is relatively air impermeable and durable can be used as an air barrier.[11] By limiting the ability of air to move through the building envelope, an air barrier will reduce convective heat loss or gain and will also reduce moisture carried by uncontrolled air infiltration.

---

[11]Keeler, Marian and Burke, Bill, *Fundamentals of Integrated Design for Sustainable Building* (Hoboken, NJ: Wiley, 2009).

Figure 8-17

HVAC Zones: A large building would require at least five thermal zones based on differences in orientation and exposure to the outside environment. Each zone will have an independent temperature control.

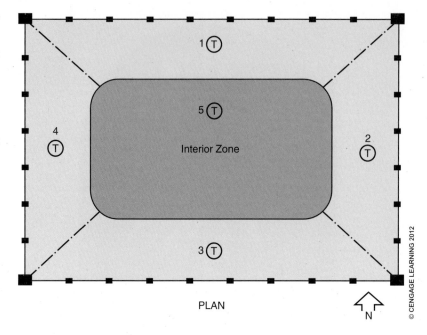

PLAN

© CENGAGE LEARNING 2012

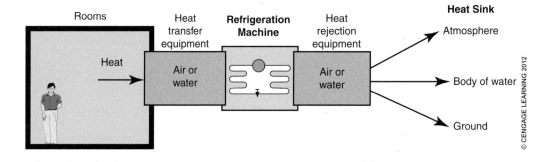

Figure 8-18

HVAC Operation: In order to cool a building by HVAC, a refrigeration system must pump heat from the interior space out into a heat sink. The heat sink can be the outdoor ambient air, body of water, or the ground. Cooling systems vary mostly by the way heat is transferred from the rooms to the refrigeration machine and from there to the heat sink.

Mechanical Systems for Heating, Cooling, and Ventilation

The type of heating, cooling, and ventilation system is a function of the size of the building, the climatic conditions, building envelope, and the load profile of the building. The heating and cooling load of some buildings will be dominated by the number of people in the building. In other buildings, the load may be dominated by heat generated by equipment and lighting. Energy-efficient appliances, lighting, computers, and equipment can significantly contribute to the reduction in cooling loads. Further, an HVAC system will be divided into zones based on room conditions and thermal exposure to the outside environment (see Figure 8-17).[12] Additional zones may also be based on the differences in mixed occupancy uses, number of occupants, and the frequency of use of various spaces. The number of zones is an important factor in

[12]Lechner, Norbert, *Heating, Cooling, Lighting: Sustainable Design Methods for Architects* (Hoboken, NJ: Wiley, 2009).

selecting the type of HVAC system. A typical HVAC system will have an air side that delivers conditioned air into the building spaces and a refrigerant or fluid side that transfers heat to or from the air side depending on the heating or cooling mode of HVAC operation (see Figure 8-18). The choice of the heat-transfer methods depends on building type and size.[13]

The air distribution system is a major component of any HVAC system. They consist of air handlers, electric motors, ductwork, air diffusers, registers and grilles, energy and humidity exchangers, and control systems. Commercial building design options for improving air distribution efficiency include (1) using variable-air-volume (VAV) systems that deliver only the volume of air needed for the actual load; (2) installing VAV diffusers for individual room temperature control; (3) increasing the size of ducting where possible to allow reductions in fan speed; (4) specifying low-face-velocity air handlers to reduce air velocity across coils thereby reducing first-costs; (5) sizing fans correctly and installing variable frequency-drive (VFD) on fan motors; and (6) consideration of displacement ventilation systems that eliminate the need for ducting by supplying air through a floor plenum and using a ceiling plenum or ceiling ducts as the return.[14]

When weather conditions are ideal, outdoor air should be used to cool the interior of buildings. Economizers serve this purpose by determining when the outside air temperature and humidity are in the same range as the required conditioned air for the building, and then intake the outside air to replace the conditioned air. Ductwork and dampers are designed so that all the return air

Defining Green

Air Distribution System

An air distribution system is a major consumer of energy that involves air handlers, electric motors, ductwork, air diffusers, registers and grilles, energy and humidity exchangers, control boxes, and its control system. According to *Green Federal Facilities*, design options for improving air distribution efficiency include (1) variable-air-volume (VAV) systems, (2) VAV diffusers, (3) low-pressure ductwork design, (4) low-face-velocity air handlers, (5) proper fan sizing with variable-frequency drive (VFD) motors, and (6) positive displacement ventilation systems.

Table 8-1
Fluorescent Light Characteristics

Lamp Type	T-12	T-8	T-5
Watts	40	32	54
Initial lumens	3,200	2,850	5,000
Efficacy (lumens/watt)	80	89	93
Lumen depreciation	10%	5%	5%

© CENGAGE LEARNING 2012

[13]Lechner, Norbert, *Heating, Cooling, Lighting: Sustainable Design Methods for Architects* (Hoboken, NJ: Wiley, 2009).

[14]BuildingGreen, *Greening Federal Facilities* (Brattleboro, VT: BuildingGreen, 2001).

Figure 8-19

LED lighting.

© ISTOCKPHOTO/JANDA75

can be exhausted from the building. Turning off major components of the HVAC system can result in significant energy savings.

Lighting Systems

Because lighting can be a significant consumer of electrical energy, a primary goal for all projects should be to reduce reliance on artificial light and maximize the use of daylighting. This effort should be an integrated strategy that combines natural light and artificial light to provide high-quality, low-energy illumination for the building's spaces. Fluorescent lighting is the best source for most lighting applications because it is more efficient than incandescent lighting. Linear and compact fluorescent lamps have a variety of color-rendering values that match incandescent and even natural daylight. Fluorescent lamp diameters are measured in 1/8-inch increments. For example, T-12s are 12/8 or 1-1/2 inch in diameter, and T-8s are 1 inch in diameter. Typical linear fluorescent lamps are compared in Table 8-1. The efficacy (lumens per watt) is higher with smaller-diameter lamps. Light-emitting diodes (LEDs) produce 29 to 45 lumens per watt compared to about 12 lumens per watt for incandescent lighting (see Figure 8-19). With a projected life of at least 50,000 hours, LEDs last 20 times longer than incandescent lighting and 2 to 3 times longer than fluorescent lights.[15]

Lighting controls are vital to reducing energy loads. They can turn lights on or off in response to the presence or absence of occupants. Daylight controls can dim or turn lights on and off to compensate for levels of natural light. Daylight-linked electrical lighting control systems have the potential to reduce lighting energy use in office buildings by as much as 50%.

Plug Loads

Plug loads are devices plugged into electrical outlets around the building that not only consume substantial energy, but also increase cooling loads due to their emission of heat. In many office buildings the largest plug loads are due to desktop

---

[15]Kibert, Charles J., *Sustainable Construction: Green Building Design and Delivery* (Hoboken, NJ: Wiley 2008).

Table 8-2
On-Site Renewable Energy Systems

Type	Advantages	Disadvantages
Solar thermal (domestic hot water and space heating)	• Least expensive and short payback period • Fewer panels are generally needed when compared to PV systems • Choice of active and passive system including batch tanks, thermo-siphon, and recirculation systems • Can supplement space heating and heat pump systems	• Moving parts (i.e., pumps) can lead to plumbing failures • Heavier panel weight requires more structural support when compared to PV systems • Higher panel tilt angle for winter water heating can present a challenge for building integrated design
Photovoltaic (PV)	• PV technologies allow for integration into building roof and façade • Lighter weight and no moving mechanical parts means low maintenance • Expected efficiency improvements in coming years	• More expensive than solar thermal and wind power generation systems • Potential net metering limitations with local utilities • High summer temperatures can reduce efficiency
Wind	• Lowest kWh cost of any renewable energy source	• Generally large, non-uniform, and unsightly • Requires significant and consistent annual wind speed

© CENGAGE LEARNING 2012

computers, typically averaging about 160 watts per unit. Electrical plug loads include printers, scanners, and copiers.[16] ENERGY STAR rated appliances and equipment can provide an energy savings of up to 10%. Controlling plug loads is a major strategy to reducing energy consumption. Occupancy-sensor-controlled outlets can be used at workstations to turn off the power to task lights, computer monitors, speakers, and other nonessential equipment when a user vacates a space.

RENEWABLE ENERGY SYSTEMS

Passive-design and energy efficiency increases the effectiveness of renewable energy in reducing the environmental impact of a building. On-site renewable energy systems include solar thermal, photovoltaic (PV) cells, and wind energy. A brief summary of each is provided in Table 8-2.

Defining Green

Net-Zero-Energy Building

A building that produces on-site energy from renewable resources that is at least as much as it consumes from total energy sources over the course of a year. This includes buildings with grid-tied solar systems that rely on power from the utility company.

---

[16]Kaneda, David; Shell, Scott; Rumsey, Peter and Fisher, Mark, "IDeAsZ² Design Facility: A Case study of a Net Zero Energy, Zero Carbon Emission Office Building", *Proceedings of Rethinking Sustainable Construction 2006*, Sarasota, FL, September 18–22.

Figure 8-20

Thermosyphon solar hot water systems move water through the system without pumps.

Figure 8-21

Photovoltaic (PV) array.

Depending on the climate zone, solar thermal (domestic hot water) systems can meet up to 85% of annual service water heating needs. Solar thermal systems need to have supplemental or back-up water heaters fueled by either gas or electric power. There are two major types: passive and active solar thermal systems. Active systems use pumps to circulate water, whereas passive systems rely on a convection loop (hot fluids rise) to circulate water through the collector and storage tank system. Also called thermosiphon systems, these systems require the storage tank to be located above the collector on the roof (see Figure 8-20). This adds weight to the roof and will require structural loading considerations. There are also direct and indirect solar thermal systems. In a direct or open-loop system, the potable water used by the building occupants runs through the solar collector to be heated and returns for direct delivery to the fixtures. In an indirect or closed-loop system, an antifreeze mixture (usually propylene glycol) runs through the solar collector and returns to a heat exchanger tank, where the heat is transferred to potable water for use by the building occupants.

Electricity can be generated from the sun using photovoltaic (PV) cells. PV cells are made from conductive materials such as silicon. When solar radiation strikes a

PV cell, electrons are released from the silicon and generate electricity. PV cells are grouped together to form modules and the modules are grouped together to form a solar PV array (see Figure 8-21). There is also mounting hardware, conductors, an inverter, and, in some cases, batteries for storage of generated electricity. An inverter is used to convert the direct current (DC) produced by the solar modules into alternating current (AC) commonly used in buildings. Disconnect switches are required to provide safety during maintenance. For grid-tied systems, a disconnect switch also isolates the individual PV system from the electrical grid, which provides safety for utility company repair personnel working on the grid in the event of a grid failure.

Grid-tied PV systems eliminate the requirement for battery storage because the building can draw power from the grid at night. During the day, excess generated electricity not consumed on site will go onto the grid and provide utility credit to the building owner. The credit can be used to offset the cost of the grid-supplied electricity used at night. Grid-tied systems also have the potential to improve the stability of the utility grid and reduce the need for the construction of new electrical power plants. Orientation of a PV array due south will usually maximize production of electricity. However, from a utility company perspective, a more western orientation is better for summer afternoon solar production. Electricity produced at 3 pm is more valuable because the overall demand for electricity is higher in the afternoon than at mid-day.

The most common forms of PV modules are crystalline and thin film (also called amorphous silicon). Crystalline silicon modules are more efficient in converting light into electricity but their cost is higher. The power output of thin-film modules is less affected by shade, and they also can be used in ways that crystalline modules cannot. Comparing cost per power generated, both types are about equal. New materials have been developed to manufacture thin-film solar cells, including technologies for depositing conductive material on a substrate, similar to ink-jet printing.[17] Building integrated photovoltaics (BIPV) incorporates PV cells into building materials, including roofing (metal, tile, and shingles), skylight glazing, window walls, building facades, and carport structures. PV systems work best when exposed to direct sunlight. Analysis of daily and seasonal solar access is vital to designing an effective on-site solar electric generating system.

ENERGY MONITORING AND MANAGEMENT SYSTEMS

"If you can not measure it, you can not improve it" (Lord Kelvin, 1824–1907). To ensure building systems operate in an energy efficient manner requires monitoring and collection of energy consumption data. Energy management systems (EMSs) are software-based programs that monitor and control energy-consuming systems to ensure that the building operates efficiently and effectively. Building systems are operated relative to scheduled occupancies and the time of day and week. EMS has traditionally managed the HVAC systems but has expanded over the years to include lighting, appliances, and office equipment. EMSs can provide as much as

---

[17]Keeler, Marian and Burke, Bill, *Fundamentals of Integrated Design for Sustainable Building* (Hoboken, NJ: Wiley, 2009).

Integrated Building Design

A process in which multiple disciplines and seemingly unrelated aspects of design are integrated in a manner that creates synergistic benefits and results that are greater than the sum of the individual components.

a 10% energy savings in the smart management of building energy systems. EMSs can also monitor air quality and carbon dioxide levels.

ENERGY TOOLS AND GUIDELINES

Prescriptive and Performance-Based Energy-Efficiency Tools

Energy-efficiency tools provide two compliance options: (1) a prescriptive compliance path based on a "cookbook" approach or (2) a performance compliance path based on a whole-building energy analysis.

Prescriptive energy tools are "recipes" that prescribe specific construction values (such as R-values for insulation in various areas of the building envelope, equipment efficiency, etc.) that are intended to achieve a particular level of energy efficiency or reduced GHG emission performance. They are also prescribed by baseline energy codes, standards, and green building rating systems as simplified or alternative options to performance-based compliance tools. Because prescriptive provisions do not require energy calculations, this approach can often save time and streamline the building code plan review compliance process. However, prescriptive measures can often restrict design options and constrain innovative design, materials, and methods of construction.

The performance-based energy approach typically set targets that must be met. The IECC and ASHRAE 90.1 contains performance-based provisions requiring an energy analysis on the "proposed design" to reduce the energy consumption as compared to a minimum-code-compliant "standard design". The analysis must compare the building thermal envelope, heating/cooling, service water heating, and lighting systems. Green building rating systems typically establish performance targets such as a required percentage of energy costs savings, or a related GHG emission reduction, as compared to a similar building constructed to baseline code requirements. LEED-NC-2009 requires a minimum 10% energy savings, with increasing levels of savings associated with points for green certification rating. The IgCC and ASHRAE 189.1 require a minimum energy use savings of approximately 30% when compared respectively to IECC-2006 and ASHRAE 90.1-2007. In all cases, an energy modeling analysis must be performed that demonstrates performance improvements.

Where fossil-fuel sources are utilized, increased energy efficiency is often closely related to reduced GHG emissions. However, regional differences may complicate this relationship. For example, some regions receive most of their power from nuclear plants, whereas others may receive theirs from hydroelectric, coal-fired, or other plants, and these plants produce widely varied amounts of GHGs. Furthermore, although nuclear plants may not generate as much GHG emissions as coal-fired plants, they have other issues, including those related to the storage of nuclear waste.

The U.S. Environmental Protection Agency (EPA) and Department of Energy (DOE) ENERGY STAR program sets guidelines for above-code residential and commercial building energy efficiency and awards labels for compliant buildings. In addition to providing guidance on building envelope and fenestration design, ENERGY STAR provides guidance on the intelligent selection of roof coverings, appliances, water heaters, space heating and cooling, and office, food service, and home electronic equipment, among others, based on their impact on energy use. Many residential green building rating programs reference the ENERGY STAR for new homes as a means of compliance, or as a means of qualifying for points for specific practices, such as providing ENERGY STAR labeled appliances. The U.S. DOE Building America program for new and existing homes is intended to promote even higher-efficiency homes than does ENERGY STAR Homes, including zero-net-energy homes.

Most green building programs require third-party verification of the actual performance of various energy system components (such as duct sealing, insulation installation, building envelope tightness, etc.). Verification of actual building energy performance may be included as a component of a building commissioning plan. LEED-NC, for example, requires the formulation and implementation of a building commissioning plan. Other energy-related facets of building commissioning, such as energy system monitoring and metering, may be awarded points in a green building rating system. For more information on building commissioning, please see Chapter 10.

Green building rating systems and programs often promote integrated design that requires a team of individuals with expertise in many diverse areas to coordinate their interests on the project, the intent being to arrive at a final design that achieves the environmental goals in each impact area (site sustainability, water, energy, material resources, etc.) in a collaborative manner. See Chapter 10 for more discussion on the integrated design and project delivery process.

High-Performance Thermal Building Envelope

A high-performance thermal building envelope that exceeds energy-code-related requirements consists of the following features:

- High R-value (thermal resistance) or low U-factor (thermal transmittance)
- Fenestration (windows, doors, and skylights) with regionally appropriate high-performance thermal and solar properties
- Reduced air infiltration (such as by means of careful location and installation of air barriers and envelope sealing and testing)
- Thermal mass walls (designed to coordinate with the specific space conditioning system and regional climate)

Performance-based designs allow trade-offs in which the increased efficiency of one component may allow another component to be less efficient, as long as the overall target energy efficiency or emissions reduction metric is met. The appropriate thermal values, location, or other applicable properties for each of the energy-efficiency components must be indicated on the plans and/or specifications.

High-Efficiency Water Heating and Space Heating and Cooling Units

Whatever power source is used, whether renewable or fossil-fuel based, high-efficiency equipment has multiple benefits. For green-power-based systems, increasing the efficiency of equipment and the corresponding reduced need for energy may contribute significantly to the feasibility and potential cost-effectiveness of on-site power generation. For fossil-fuel-based systems, increased energy efficiency still means a reduction, though not the total elimination, of fossil-fuel use, as well as a corresponding decrease in GHG emissions.

Globally, more energy is used to heat water than for any other human-related purpose. Reducing the amount of water usage indirectly reduces energy demand and consumption impacts energy use. For example, if less hot water is used for bathing and laundry purposes, it also means that less energy is used.

Green rating systems, codes, and standards require equipment to be properly sized, referencing nationally recognized design standards. Improperly sized equipment, whether undersized or oversized, may not provide for comfort nor produce expected energy savings, emissions reductions, and cost savings. There are also requirements for zero- or reduced-chlorofluorocarbon (CFC)-based refrigerants for heat pumps and air conditioning systems.

For water heating and space heating or cooling equipment, equipment efficiency should be specified in accordance with the prescriptive or performance-based energy strategy. Product data sheets indicating equipment energy efficiency may also be required. Typical examples of high-efficiency water heating and space heating and cooling units include:

- High-efficiency natural or fuel-gas-based heating units
- Solar-based or solar-assisted units
- Air-to-air heat pumps
- Water-to-air (geothermal) heat pumps
- Other, especially commercial, examples

Efficient Duct System Design, Air Flow Balancing, Duct Sealing, and Duct Leakage

Properly designed, balanced, and sealed duct systems can reduce the thermal losses inherent in the process of moving conditioned air from the heating or cooling unit to the spaces that require conditioning. They also ensure that the proper amount of conditioned air is supplied to each space. Energy and mechanical code referenced standards utilized for duct system design, as well as requirements for air flow balancing, duct sealing, and duct leakage, must be provided in the construction documents. Duct sizing calculations and third-party verification of duct leakage or inspection of duct sealing are also required by baseline energy codes and standards.

Ceiling Fans, Whole-Building Ventilation, and Natural Ventilation

As mentioned earlier in this chapter, air movement can have a significant effect on human comfort, generally making higher temperatures and humidity levels more tolerable for building occupants, as well as moving air of desirable temperatures to areas where occupants are located. Ceiling fans can be effective in both residential and commercial applications.

Natural ventilation is accepted as an alternative to the mechanical ventilation requirements of the building and mechanical codes provided they meet the prescriptive requirements of the codes or acceptable standards. Operable windows with insect screens that are intended to provide natural ventilation should be indicated on the plans.

Whole-house or whole-building ventilation systems function much like manual versions of demand-controlled ventilation systems or economizers (see following discussion). They are typically used in conjunction with natural ventilation (operable windows) to draw comfortable outdoor air (such as cooler night air or air from low window openings on the cooler sides of a building on a warm day) into building interiors, sometimes up through a ventilated attic, and ultimately to the exterior, and are typically driven by means of a fan that is centrally located in the ceiling of the upper floor in a building of small to moderate size.

Ceiling fans and natural ventilation systems, when used intelligently (to augment rather than combat the function of a mechanical space conditioning system), can be a very effective means to conserve energy. The effectiveness of natural ventilation may be increased by proper location of operable openings, such as openings to a covered porch or sun space.

Economizers and Demand-Controlled Ventilation (DCV)

As previously mentioned, economizers save cooling costs by detecting when the outdoor air temperature is in a comfortable range to be used for building cooling purposes. Some economizers also detect outdoor air humidity (enthalpy control). Demand-controlled ventilation systems save energy by reducing the amount of outdoor air that must be conditioned through monitoring of the actual indoor air quality (CO_2 and/or humidity, etc.) and automatically introducing outdoor air only when it has advantages, as opposed to always delivering a fixed amount of outside air.

For economizers, energy recovery systems, and demand-controlled ventilation systems, energy/mechanical codes and standards provide performance-based and prescriptive-based requirements. Economizer and demand-controlled ventilation system details must be provided on the construction documents.

Energy-Efficient Appliances and Equipment

Select energy-efficient appliances such as ENERGY STAR labeled appliances for both residential and commercial use. Specifications should indicate which appliances, lighting fixtures, ceiling fans, and so forth are to be ENERGY STAR or equivalent, along with proof of labeling or equivalency. Evidence of labeling or equivalency is provided by means of product data sheets and field verification. Locations of ceiling fans and whole-house or whole-building ventilation equipment, as well as other ventilation system details, must be indicated on the plans and installation must be field verified.

Energy-Efficient Lighting Strategies

The use of natural light to illuminate interior spaces during daytime hours reduces the energy that might otherwise be required to artificially illuminate those spaces. However, careful attention to inevitable compromises in building envelope thermal performance must also be carefully considered (i.e., the comparison of heating and cooling load increases versus lighting power load decreases). In addition to energy implications, daylighting and views can have a positive impact on human comfort, health, sense of well-being, and productivity.[18] Once again, however, these benefits should be carefully weighed against possible discomfort that may arise from hot or cool zones and glare, as may occur adjacent to large glazed areas in the building envelope, and the ability of the space heating and cooling system to effectively compensate for differences in conditioning requirements for these specific portions of the building. The effect of the sun seasonally and at various times of day, for example, will have various impacts for areas adjacent to glazed openings on each side of the building, or for shaded versus non-shaded openings. In most cases, these negative potential effects can be eliminated or mitigated by means of shading devices, which may sometimes take the form of simple window treatments.

For daylighting, building codes provide prescriptive-based minimum requirements for natural lighting, but allow artificial to substitute for natural lighting altogether. Most green building rating systems contain provisions that award points for daylighting.

Providing maximum lighting (illuminance) levels for specific tasks, spaces, or building types reduce lighting energy use. Maximum lighting power (or lighting power density) for building interiors and exteriors is regulated by energy codes and standards.

Energy-efficient light fixtures and lamps—such as those that employ energy fluorescent, compact florescent (CFL), or light-emitting diode (LED) technologies—can save substantial amounts of energy over other less efficient fixtures and lamps, such as incandescent bulbs. Note that ENERGY STAR labels both light fixtures and lamps. For energy-efficient lighting fixtures (such as CFLs and LEDs), green building rating systems often contain provisions that specifically encourage their use by awarding points.

Energy use can be reduced by providing artificial light only when it is necessary. For occupancy or motion detectors, energy codes and standards contain requirements and green building rating systems contain provisions that encourage their use. Solar-powered light fixtures do not draw power from the grid and reduce the load on on-site green-power-generation equipment. For solar-powered lighting fixtures, green building rating systems contain provisions that encourage their use. Energy codes indirectly encourage their use through exterior lighting power maximum requirements.

The locations, and possibly the number or percentage, of energy-efficient luminaries, occupancy/motion detectors, and solar-powered light fixtures should be indicated in the construction documents.

---

[16]Guzowski, Mary, *Daylighting for Sustainable Design* (New York: McGraw-Hill, 2000).

GREEN BUILDING RATING SYSTEMS

LEED-NC

LEED-NC is a rating system composed primarily of elective provisions that are assigned points. LEED-NC contains few mandatory provisions. Its mandatory provisions are its "prerequisites." The intent of LEED-NC's Energy and Atmosphere section is to reduce the amount of energy used in buildings and promote the use of renewable energy production, thereby reducing the negative environmental consequences of fossil-fuel-based energy generation. There are three prerequisite and six optional credits (see Table 8-3).

Green Globes

Green Globes is a rating system containing elective provisions that are assigned point values. Green Globes does not contain mandatory requirements. The Green Globes Energy category assessment consists of online survey questions organized under the following six stages of the design construction process: (1) Pre-design, (2) Schematic Design, (3) Design Development, (4) Construction Documents, (5) Contracting/Construction, and (6) Commissioning. Each stage assesses energy-efficiency priorities with the objective of reducing environmental impacts (see Table 8-4).

LEED for Homes

LEED for Homes is a rating system composed primarily of elective provisions that are assigned points. LEED-NC contains few mandatory provisions. Its mandatory provisions are its "prerequisites." Similar to LEED-NC, the intent of LEED for Home's Energy and Atmosphere section is to reduce the amount of energy used in buildings and promote the use of renewable energy production, thereby reducing

Table 8-3
LEED for New Construction: Energy and Atmosphere

Credit	Prerequisite and Requirements
EAp1	Fundamental Commissioning of Building Energy Systems
EAp2	Minimum Energy Performance
EAp3	Fundamental Refrigerant Management
EAc1	Optimize Energy Performance
EAc2	On-site Renewable Energy
EAc3	Enhanced Commissioning
EAc4	Enhanced Refrigerant Management
EAc5	Measurement and Verification
EAc6	Green Power

Table 8-4
Green Globes Energy Assessment by Project Stage

Pre-Design Stage
- Objective to establish an energy target
- Objective to minimize building energy demand
- Objective to integrate energy-efficient systems
- Objective to integrate renewable energy sources
- Objective for energy-efficient transportation

Schematic Design Stage
- Modeling and simulation of building energy performance: establishing an energy target
- Objective to minimize building energy demand

Design Development Stage
- Modeling and simulation of energy use
- Energy demand minimization strategies

Construction Documents Stage
- Building energy performance

Contracting and Construction
- Energy efficiency
- Renewable sources of energy

Commissioning
- Energy target
- Commissioning to ensure minimization of building's energy demand
- Commissioning of energy-efficient systems
- Commissioning of renewable energy source systems

the negative environmental consequences of fossil-fuel-based energy generation. Project must meet ENERGY STAR for Homes v2 requirements. There are two pathways to do so, either by energy modeling or through a prescriptive list of measures. An additional prerequisite must be met, Residential Refrigerant Management. The energy modeling pathway has 2 prerequisites and 3 credits. The prescriptive pathway has 6 prerequisites and 10 optional credits (see Table 8-5).

CODES AND STANDARDS

GBI-01 Green Building Assessment Protocol for Commercial Buildings

GBI-01 is a rating system and an ANSI-approved standard. Chapter 8 of GBI-01 regulates energy. GBI-01 requires that at least 150 points be earned to qualify for its Path A performance option, or 100 points be earned to qualify for its Path B prescriptive option. There are 300 points available in the performance option and 250 points available in the prescriptive option. See Table 8.6 for a summary of the energy provisions of GBI-01.

Table 8-5
LEED for Homes: Energy and Atmosphere

Credit	Prerequisites and Requirements
EA1	Optimize energy performance – performance
EA2	Insulation
EA3	Air infiltration
EA4	Windows
EA5	Duct tightness
EA6	Space heating and cooling
EA7	Water heating – both
EA8	Lighting
EA9	Appliances
EA10	Renewable electric energy generation
EA11	Residential refrigerant management – both

ICC 700 National Green Building Standard

ICC 700 is a rating system and an ANSI-approved standard. Chapter 7 of ICC 700 addresses energy efficiency. Although Chapter 7 of ICC 700 primarily contains elective provisions, it contains mandatory provisions that are applicable to both its prescriptive- and performance-based paths. In addition, the standard requires increased points in energy at each of its performance levels. A minimum of 30 points at the Bronze, 60 points at the Silver, 100 points at the Gold, and 120 points at the Emerald performance level must be acquired from the energy efficiency provisions of ICC Chapter 7. See Table 8.7 for a summary of these provisions as they relate to new buildings.

ASHRAE 189.1 Standard for the Design of High-Performance Green Buildings

The intent of the energy-efficiency section of ASHRAE 189.1 is to significantly reduce the energy consumption of new buildings by energy loss reductions through the building envelope, increased mechanical system efficiencies, reduced lighting loads, and other energy-saving measures. ASHRAE 189.1 includes a set of mandatory requirements that must be met for all projects, and an option for either a prescriptive or performance set of requirements. Buildings complying with ASHRAE 189.1 achieve an overall average of 30% energy-use savings compared to those complying with ASHRAE 90.1-2007.

Prescriptive Envelope Requirements

Insulation and Fenestration: The prescriptive building envelope requirements in Standard 189.1 include energy-saving measures that are above Standard 90.1. The

Table 8-6
GBI-01 2010 Green Building Assessment Protocol for Commercial Buildings

Chapter 8: Energy

8.1 Building Carbon Dioxide Equivalent (CO_2e) Emissions—PATH A (Performance Option)

8.1.1 Percent Reduction in Carbon Dioxide Equivalent (CO_2e) Emissions

150 points are awarded for a 50% reduction in CO_2e emissions, with an additional point awarded for every 1% reduction, up to a maximum of 250 points. The reduction is calculated in accordance with formulas that establish a relationship between a building with a baseline equivalent emission rate and the equivalent emission rate for the proposed design.

- For the base building, EPA's Target Finder is used to determine energy use intensity (EUI), and the baseline building's site EUI must be 50% better than a Target Finder score of 50.
- For the proposed design, EUI is determined by calculations in accordance with Section 8.1.1 and computer-based simulation programs that conform to either Section 506 of the 2009 International Energy Conservation Code or Section G2.2 of Appendix G to ASHRAE 90.1-2007.
- The CO emission factor for each fuel in the baseline building's and the proposed building's annual energy fuel mix is determine by Table 8.1.1.

8.2 Demand—PATH B (Prescriptive Option)

8.2.1 Passive Demand Reduction

8.2.1.1
- 4 points awarded for thermal mass walls in Climate Zones 1 through 5 where at least 20% of the building envelope complies with either of the following:
    a) Have a heat capacity of 7 Btu/ft^2 ^{o}F (143kJ/m^{2}K) or more, **or**
    b) Where the walls have a material unit weight of not more than 120 lb/ft^3, have a heat capacity of
       5 Btu/ft^2 ^{o}F (143kJ/m^{2}K)

8.2.1.2
- 4 points awarded for thermal mass walls in Climate Zones 1 through 5 where at least 20% gross area of walls used as interior partitions are not provided with an insulating material or wallboard as an interior finish and comply with either of the following:
    a) Have a heat capacity of 7 Btu/ft^2 ^{o}F (143kJ/m^{2}K) or more, **or**
    b) Where the walls have a material unit weight of not more than 120 lb/ft^3, have a heat capacity of
       5 Btu/ft^2 ^{o}F (143kJ/m^{2}K)

8.2.1.3
- 4 points awarded for thermal mass floors in Climate Zones 1 through 5 where at least 50% of return air plenums are located directly in contact with floors that either:
    a) Have a heat capacity of 7 Btu/ft^2 ^{o}F (143kJ/m^{2}K) or more, **or**
    b) Where the floors have a material unit weight of not more than 120 lb/ft^3, have a heat capacity of
       5 Btu/ft^2 ^{o}F (143kJ/m^{2}K)

8.2.2 Thermal Energy Storage System

8.2.2.1
- 4 to 12 points awarded for buildings in Climate Zones 1 through 5 where a thermal energy storage system is used to offset the peak cooling demand by more than 30% to more than 50%.

8.2.3 Power Demand Reduction

8.2.3.1
- 4 points awarded where more than the building's average monthly power demand factor is greater than 0.75, or 8 points where it is greater than 0.85.

Table 8-6 *(Continued)*

8.2.4 Demand Capable Energy Management System
- 4 points awarded where an energy management system reduces power below the non-reduced peak by at least 15%, or power demand is controlled by the electric utility based on a 15% load shedding agreement, or 8 points where these percentages are increased to 30%.

8.3 Measurement and Verification

8.3.1 Measurement Verification Protocol

8.3.1.1
- 8 points awarded where an Energy Metering Reporting Plan in the Operations and Maintenance manual is provided for buildings with an area of over 20,000 sq. ft. and the requirements of this section are satisfied.

8.3.1.2
- 2 points awarded for buildings with an area of less than 20,000 sq. ft. where a Measurement and Verification program is implemented in accordance with this section.

8.4 Building Opaque Envelope

8.4.1 Thermal Resistance and Transmittance

8.4.1.1
- 12 points awarded where all opaque elements of the building envelope comply with the thermal resistance and thermal transmittance requirements of this section.

8.4.2 Orientation

8.4.2.1
- 1 to 6 points awarded where the building is oriented such that the ratio of north/south fenestration area to east/west fenestration area is at least 1.25 to 2.0.

8.4.3 Fenestration Systems

8.4.3.1
- 12 points awarded where fenestration thermal transmittance complies with Table 8.4.3-A.

8.4.3.2
- 12 points awarded where building fenestration system Solar Heat Gain Coefficient (SHGC) complies with Table 8.4.3-A.
- 2 points awarded where the design professional documents how the building uses assemblies that perform multiple functions.

8.5.1 Daylighting

8.5.1.1
- 1 to 8 points awarded where combined sidelight and toplight daylight areas is at least 10 to at least 50% of the net building area.

8.5.1.2
- 4 points awarded where buildings in Climate Zones 1, 2, 3A or 3B have a minimum effective aperture for vertical fenestration of at least 0.10_{vf} or buildings located in Climate Zones 3C, 4, 5, 6, 7 or 8 have an effective aperture for vertical fenestration of at least 0.15_{vf}.

8.5.1.3
- 3 points awarded where skylights comprise 2 to 6% of the roof area in Climate Zones other than 7 and 8.

8.6 HVAC Systems and Controls

8.6.1 Cooling Equipment

8.6.1.1
- 1 to 5 points awarded relative to the Cooling Equipment Base Efficiencies indicated in Table 8.6.1.1.

(Continues)

Table 8-6 (*Continued*)

8.6.1.2
- 1 to 10 points awarded for cooling equipment incremental efficiency improvements in accordance with Table 8.6.1.2.

8.6.2 Cooling Towers
- 3 points awarded where cooling tower fan energy is reduced by the use of two speed fans, variable speed fans or other measures.

8.6.2.2
- 3 points awarded where a waterside economizer system is installed in accordance with this section.

8.6.3 Heating Equipment

8.6.4.1
- 1 to 12 points awarded where heating equipment annual fuel utilization efficiency, thermal efficiency or combustion efficiency is in accordance with this section.

8.6.5 Condensate Recovery

8.6.5.1
- 1 to 3 points awarded where steam heating systems, including district systems, are equipped to recover and return condensate in accordance with this section.

8.6.6 Steam Traps

8.6.6.1
- 2 points awarded where all steam traps are sealed/stamped by a Professional Engineer and isolation valves are installed to accommodate repairs.

8.6.7 Domestic Hot Water Heaters

8.7.1
- 2 points awarded where all domestic hot water heaters comply with the efficiency requirements of ASHRAE 90.1-2007 or the 2009 International Energy Conservation code and are intermittent electric igniters and low NO_x burners are provided.

8.6.8 Variable Speed Control of Pumps

8.6.8.1
- 1 to 5 points awarded where variable speed control of pumps is provided for at least 15% to at least 75% of hydronic pumps greater than 3 hp.

8.6.9 Minimizing Reheat and Re-cool

8.6.9.1
- 4 to 8 points awarded for the use of central multiple zone HVAC systems or other strategies that reduce or eliminate reheat and re-cool.

8.6.10 Air Economizers

8.6.10.1
- 1 point awarded where air economizers have a mode that uses outdoor air for cooling.

8.6.11.2
- 1 point awarded where flexible duct work is installed in accordance with this section.

8.6.11.3
- 1 point awarded where sealed duct joints and seams are leak-tested in accordance with this section.

8.6.11.4
- 1 point awarded where fan motors meet the efficiency requirements of this section.

8.6.11.5
- 2 points awarded where variable speed fans are controlled in accordance with this sections.

Table 8-6 (*Continued*)

8.6.12 Demand Controlled Ventilation

8.6.12.1
- 5 points awarded where ventilation rates are controlled using occupancy or CO_2 sensors.

8.6.12.2
- 5 points awarded where ventilation heat recovery is used on systems listed in this section where they are not required to be provided with ventilation heat recovery by ASHRAE 90.1-2007 or the 2009 International Energy Conservation Code.

8.7 Lighting Systems and Controls

8.7.1 Total Lighting Power Density

8.7.1.1
- 13 points awarded where total lighting power density is in accordance with Tables 8.7.1-A or 8.7.1-B.

8.7.2 Interior Automatic Light Shutoff Controls

8.7.2.1
- 3 points awarded where buildings with an area of over 5,000 sq. ft. are provided with time-scheduling or individual occupant-sensing devices in accordance with this section.

8.7.3 Light Reduction Controls

8.7.3.1
- 7 points awarded where the lighting load is reduced by at least 50% and controls are provided in accordance with this section.

8.7.4 Controls for Daylight Zones

8.7.4.1
- 2 to 8 points awarded where each sidelight or toplight daylight area that is between 250 and 2,500 sq. ft. is provided with manual or automatic controls in accordance with this section.

8.7.4.2
- 2 to 6 points awarded where sidelight and toplight daylight areas greater than 2,500 sq. ft. are provided with automatic control strategies in accordance with this section.

8.7.5 Exterior Lighting Controls

8.7.5.1
- 3 points awarded where controls are installed for exterior lighting in accordance with this section.

8.7.6 Exterior Luminaries

8.7.6.1
- 4 points awarded where exterior luminaires are installed in accordance with this section.

8.7.6.2
- 2 points awarded where all exterior lighting is provided by pulse-start metal halide lamps.

8.8.1 Elevators and Escalators

8.8.1.1
- 3 points awarded where regenerative breaking installed.

8.8.1.2
- 2 points awarded where moving walkways and escalators have the capability to slow down or stop when detectors indicate no traffic.

8.9 Renewable Energy

8.9.1
- 1 to 50 points awarded for a commitment to purchase green electrical power or certified renewable energy certificates (RECs) in accordance with this section.

prescriptive requirements can only be used when vertical fenestration is less than 40% of gross wall area. When the vertical fenestration is greater, the building must comply with the performance requirements.

Fenestration Shading: The standard requires permanent shading projections on the west, south, and east walls of buildings in Climate Zones 1 through 5. There is a requirement is for an area-weighted projection factor of 0.5. Requirements can also be met by building configurations and architectural elements.

Orientation: The prescriptive fenestration requirement for Climate Zones 1 through 6 encourages design that orients the building with the longest facades on

Table 8-7
ICC 700-2008 National Green Building Standard

Chapter 7—Energy Efficiency
Minimum Energy Efficiency Requirements
Buildings can follow either the prescriptive- or performance-based paths, but must always comply with the mandatory provisions of Section 701.4
ENERGY STAR Qualified homes are deemed to achieve the Bronze level for energy performance in Chapter 7.
The performance-based compliance path must be used to demonstrate compliance for the Emerald level.
Compliance with the energy provisions of Chapter 7 must be reviewed by the Adopting Entity or a designated third party.
Mandatory Practices
HVAC systems must be sized in accordance with ACCA Manual J.
Industry-approved guidelines must be used to design radiant or hydronic space heating systems.
Duct systems must be sealed in accordance with 701.4.2.1.
Building cavities are not permitted to be used as supply ducts.
Penetrations between conditioned and unconditioned space must be fully sealed in accordance with 701.4.3 Item 2.
Insulation in floors, foundations and crawlspaces must be installed in accordance with Section 701.4.3.2.
Window and door openings must be installed to form a complete air barrier, band and rim joists and the joint between foundations and sill or bottom plates must be insulated and sealed, skylight shafts and knee walls must be insulated, and building envelope insulation and air sealing at exterior walls common to decks and stairs must be in accordance with 701.4.3.3.
Attic access knee walls, doors and eave vents and recessed lighting fixtures that penetrate the building thermal envelope must be in accordance with 701.4.3.4.
Fenestration U-Factors and SHGC must be in accordance with 701.4.4.1.
Performance Path
Points from the prescriptive path cannot be combined with points from the performance path.
Points are awarded for performance based compliance as follows:
• 30 points awarded where energy cost performance exceeds that of the 2006 IECC by at least 15%.
• 60 points awarded where energy cost performance exceeds that of the 2006 IECC by at least 30%.
• 100 points are awarded where energy cost performance exceeds that of the 2006 IECC by at least 50%.
• 120 points are awarded where energy cost performance exceeds that of the 2006 IECC by at least 60%.
Compliance must be determined in accordance with the requirements of the 2006 IECC.

Table 8-7 (*Continued*)

Prescriptive Path

10 to 36 points are awarded where total building thermal envelope UA is improved by 10 or 20% in accordance with Table 703.1.

10 or 15 points are awarded for Grade 2 or Grade 1 insulation installation, respectively, where compliance with 703.1.2 through 703.1.2.4 is verified by a third party.

3 points awarded where no third-party verification is performed, or 15 points awarded where their-party verification is performed, and insulation and air sealing is in accordance with 703.2.1.1.1 through 703.2.1.4.

5 to 8 points awarded where building fenestration U-Factors and SHGC are in accordance with 703.3.1.

4 points awarded where a combination space heating and water heating system is provided in accordance with 703.4.1.

1 to 17 points awarded where furnace and boiler efficiencies are in accordance with 703.4.2.

1 to 18 points awarded where cooling equipment efficiency is in accordance with 703.4.5.

20 to 30 points are awarded where ground source heat pumps are provided in accordance with 703.4.6.

1 point awarded where all ceiling fans are in accordance with 703.4.7.

2 points awarded where whole building or whole building dwelling unit fans are provided in accordance with 703.4.8.

1 to 4 points awarded where an advanced electric and fossil fuel submetering system is installed in multi-unit buildings in accordance with 703.4.9.

1 to 10 points awarded where water heating energy efficiency is in accordance with 703.5.1.

2 points awarded where drain-water heat recovery is provided in a multi-family unit building in accordance with 703.5.3.

1 point awarded where hot water lines are insulated in accordance with 703.5.4.1.

1 point awarded where boiler supply piping in unconditioned spaces is insulated.

1 point awarded where indirect-fired water heater storage tanks are heated in accordance with 703.5.5.

Additional Practices

Points for additional practices can be added to both the prescriptive- and performance-based energy compliance paths.

4 to 8 points are awarded where lighting and appliances are in accordance with 704.2.1.

2 points are awarded where recessed light fixtures that penetrate the building thermal envelope are in accordance with 704.2.2.

2 to 4 points are awarded where occupancy sensors are provided in accordance with 704.2.3.

2 points awarded where tubular daylighting devices or skylights are in accordance with 704.2.4.

2 to 5 points awarded for each ENERGY STAR refrigerator, dishwasher or washing machine in accordance with 704.2.5.

1 point awarded where an induction cooktop is provided.

1 point awarded where occupancy sensors are installed in accordance with 704.2.7.

5 points awarded where sun-tempered design practice that is in accordance with 704.3.1.1.

1 point awarded where windows are shaded in accordance with 704.3.1.2.

1 point per item is awarded for each passive design feature that complies with 704.3.1.3.

4 points awarded where a passive solar heating design is provided in accordance with 704.3.1.4.

8 to 20 points are awarded where a solar water heater heating system is provided in accordance with 704.3.2.1.

1 point for each 1/10 KW is awarded where photovoltaic panels are provided in accordance with 704.3.3.1.

½ point awarded for each 1/10 kW produced by on-site renewable energy sources other than photovoltaic panels in accordance with 704.3.3.2.

5 points awarded where duct systems are designed and installed in accordance with 704.4.1.

15 points awarded where a ducts are not provided in a space heating system in accordance with 704.4.2.

15 points awarded where ducts are not provided in a space cooling system in accordance with 704.4.3.

(*Continues*)

Table 8-7 *(Continued)*

Additional Practices
12 points awarded where ductwork is in accordance with 704.4.
5 points awarded where return ducts or transfer grilles are installed in rooms with doors in accordance with 704.4.5.
1 point awarded where heating or cooling equipment is selected in accordance with 704.5.1.
1 point awarded where HVAC contractor and service technician are certified in accordance with 704.5.2.
3 points awarded where heating or cooling system performance is verified by the HVAC contractor in accordance with 704.5.3.
4 points awarded where air handler leakage and testing is in accordance with 704.5.5.
5 points awarded where third-party on-site inspection is conducted to verify compliance of energy system features in accordance with 704.6.1.
3 to 15 points awarded where building envelope leakage, whole building ventilation, fossil fuel furnaces and water heaters, fireplaces and fuel burning appliances are in accordance with 704.6.2.1.
5 to 15 points awarded where HVAC duct systems are tested for air leakage in accordance with 704.6.2.2.
8 points awarded where HVAC airflows are verified in accordance with 704.6.2.3.

Innovative Practices
2 to 7 points awarded where devices are installed to monitor energy consumption in accordance with 705.1.
2 to 5 points awarded where a renewable energy service plan is provided in accordance with 705.2.

the north and south. This minimizes fenestration on the east and west facades that are subject to solar gains during the morning and afternoon of the summer months.

On-site Renewable Energy

ASHRAE 189.1 has a mandatory requirement for buildings to provide space for the future installation of a photovoltaic, solar thermal, geothermal energy, or wind system with a minimum rating of 13 Btu/h·ft^2 (40 W/m^2) multiplied by the total roof area. This ensures space for solar collectors, pathways for conduit and piping, and associated equipment. Under the prescriptive path there is a requirement to install on-site renewable energy systems at the time of construction. These systems must be capable of producing at least 6 kBtu/ft^2 (20 kWh/m^2) on an annual basis. Sites that meet the poor incident solar radiation exception must purchase a specified amount of renewable energy credits (RECs) complying with the Green-e Energy National Standard for Renewable Electricity Products.

Mechanical Equipment

ASHRAE 189.1 encourages the installation of equipment with higher efficiencies than the minimum set by the National Appliance Energy Conservation Act (NAECA), the 1992 EP Act, and the Energy Independence and Security Act (EICA). Two options exist for compliance with the prescriptive path. The first option is to comply with the minimum efficiency requirements of NAECA, EP Act, and EICA and the renewable energy and peak load reduction requirements as outlined in the prescriptive portion of the standard. The second option allows a reduction by one-third in the amount of on-site renewable energy required and a less stringent peak load reduction if the mechanical equipment efficiencies are increased to meet ENERGY STAR requirements and the requirements in Appendix C of ASHRAE 189.1.

Along with increased duct sealing requirements, demand control ventilation is required to limit the amount of outside air ventilation based on the number of occupants in a room or space. In addition to expected energy savings, this will help avoid mold and associated moisture problems, resulting in a healthier and more energy-efficient building. The standard provides higher efficiency requirements for exhaust air energy recovery devices that transfer heat from the exhaust air to the intake air for heating during the winter months and transfer heat from the intake air to the exhaust air for cooling during the summer months.

To minimize energy waste associated with simultaneous operation of heating and cooling equipment, zone controls are required based on amount of supply air flow. These controls are intended to prevent reheating, recooling, and the mixing or simultaneous supply of air. In addition, the standard reduces the allowable fan power by 10% by requiring at least a two-speed fan or a variable-speed fan that allows for reductions in fan power at lower loads.

To reduce energy in unoccupied hotels and motel guest rooms, controls must be installed that set back the HVAC system and turn off plug loads, including lighting, switched outlets, and televisions, when the rooms are unoccupied.

Energy-Consumption Data Collection

ASHRAE 189.1 requires that measurement devices be installed to collect energy-consumption data. Depending on electrical load sizes, submetering and data collection devices must be installed on energy subsystems such as HVAC systems, water heating, lighting, process equipment, plug loads, etc. The data storage system must be capable of producing reports summarizing the data so that building performance can be assessed on at least a monthly basis. Daily data with hourly energy use profiles must be retained for at least three years.

Service Water Heating

ASHRAE 189.1 requires more piping insulation and higher efficiency ratings for water heaters. In addition, insulation is required on the sides and bottoms of heated pools used primarily as spas or for physical therapy.

Peak Load Reduction

ASHRAE 189.1 requires that buildings contain automatic systems for peak load reduction that are capable of reducing the electrical peak demand of the building by not less than 10%. Automatic peak load reduction can be accomplished through demand limiting, in which load use is curtailed or reduced or through load shifting in which the load use is moved to off-peak hours.

Lighting

ASHRAE 189.1 requires a 10% reduction in energy used for lighting as compared to the ASHRAE 90.1-2007 energy standard. Occupancy sensors, automatic controls for egress and security lighting, daylight sensor controls, and controls for outdoor lighting are required. Automatic controls must be installed to reduce all outdoor lighting by a minimum of 50% within an hour after normal business closing.

Performance Option

The performance compliance path requires a whole-building energy-simulation capable of modeling annual energy use on an hourly basis. The simulation must demonstrate that the annual building energy cost of the proposed building is less than or equal to the annual building energy costs of a baseline comparison building that is designed to meet the mandatory and prescriptive requirements. In addition, the annual CO_2-equivalent emissions (carbon footprint) of the proposed building's energy use must be less than or equal to the CO_2-equivalent emissions level of the baseline comparison building.

International Green Construction Code (IgCC)

The IgCC is a code composed primarily of minimum mandatory requirements. As mentioned in previous chapters, the International Green Construction Code (IgCC) provides a comprehensive set of requirements intended to reduce the negative impact of buildings on the natural environment. Chapter 6 of the IgCC contains requirements for energy conservation, efficiency, and atmospheric quality.

Unlike the energy requirements of most green building standards and rating systems, the energy chapter of the IgCC stands on its own and does not defer to other energy codes or standards for most of its detailed design and construction requirements.

All of the provisions in the IgCC discussed in Table 8-8 are mandatory except where they are indicated to be project electives. Project electives are selected from IgCC Table 303.1 by the owner or designer and may vary from project to project. The total number of project electives required on each project is determined by the jurisdiction in Table 302.1. The energy related project electives encourage and recognize energy performance that is beyond the minimum requirements of the IgCC. For a summary of the IgCC's energy provisions see Table 8-8.

THE IMPORTANCE OF ENERGY AND APPROPRIATE BUILDING DESIGN

When the extraction, production, transmission, and consumption of nonrenewable energy resources such as fossil fuels and nuclear energy are considered, energy has the greatest impact environmental impact. Based on specific climate region and site conditions, there are always constraints and opportunities to create a low energy profile while tapping into regional renewable energy resources. Energy efficient and conserving building design must always first strive to reduce energy load through passive design strategies, then selecting energy-efficient HVAC, lighting, and water heating systems, and lastly utilize on-site renewable energy.

Table 8-8
IgCC Mandatory Energy Provisions (Public Version 2.0)

Chapter 6: Energy Conservation, Efficiency and Atmospheric Quality

Section 602: Energy Performance, Peak Power and Reduced CO_2e Emissions

602.1 Zero energy performance index (zEPI). The building shall be designed and constructed to have a zEPI not greater than 51. The zEPI shall be calculated in accordance with Section 603.1.1. Buildings complying with the 2006 International Energy Conservation Code shall be deemed to have a zEPI of 73.

602.2 Compliance paths. Buildings with an aggregate area of over 25,000 square feet are required to use the IGCC's *performance*-based or outcome-based energy compliance path. Buildings with an aggregate area of 25,000 square feet or less may use either the *prescriptive-based, performance-based,* o*utcome-based* or *energy use intensity (EUI) compliance path* of the IGCC.

602.3 Documentation and verification for alterations to existing buildings. The energy performance associated with alterations to existing buildings shall be documented by the building owner in accordance with this section.

Section 603: Energy Use and Atmospheric Impacts (*Performance and Outcome-based Compliance Path*)

603.1 Determination of building annual energy use, peak energy demand and reduced CO2e emissions. Where buildings are designed using the performance-based compliance path, the methods for calculating and verifying annual energy use, peak energy demand, and reduced CO_2e emissions shall be in accordance with this section.

603.2 Determination of and compliance with building annual net energy performance, peak net energy demand and CO_2e emissions requirements. Where buildings are designed using the outcome-based compliance path, the annual net energy performance, peak net energy demand, and annual direct and indirect CO2e emissions shall meet the provisions of Table 603.2(1) through 603.2(3). Annual net energy performance and peak net energy demand shall be based on the estimated energy use of the building for all purposes for 12 continuous months prior to commissioning and based on actual utility consumption for 12 continuous months after commissioning.

603.3 Calculation procedures. The annual energy use of the building and building site shall be calculated in accordance with this section.

603.4 Qualified software for determinations of annual energy use. Calculation software tools and procedures used to comply shall include the capabilities identified in accordance with Section 506.6 of the International Energy Conservation Code.

603.5 Design professional in responsible charge of building energy simulation. Where it is required that documents be prepared by a registered design professional, the code official shall be authorized to require the owner to engage and designate on the building permit application a registered design professional who shall act as the registered design professional in responsible charge of building energy simulation. The code official shall be notified in writing by the owner if the registered design professional in responsible charge of building energy simulation is changed or is unable to continue to perform the duties.

603.6 Minimum requirements for buildings pursuing performance compliance path. Buildings following the performance-based compliance path shall meet the minimum requirements of a code compliant standard reference building design without regard to technology choice in the proposed design.

Section 604: Energy Metering, Monitoring and Reporting (*Prescriptive and Performance Compliance Path*)

604.3 Energy distribution design requirements and load type isolation. Energy distribution systems within, on or adjacent to and serving a building shall be designed such that each primary circuit, panel, feeder, piping system or supply mechanism supplies only one energy use type as specified. The energy distribution system shall be designed to facilitate the collection of data for each of the following building energy load categories.

- HVAC system total energy (including, but not limited to, fans, pumps, boiler energy, chiller energy and hot water)
- Lighting system total energy use (all interior and exterior lighting)

(Continues)

Table 8-8 (Continued)

- Energy used for building operations (vertical transportation systems, automatic doors, motorized shading systems, ornamental fountains and fireplaces, swimming pools, snow-melt systems and other similar operations)
- Plug loads (devices, appliances and equipment connected to convenience receptacle outlets)
- Process loads (associated with activities within the building such as, but not limited to, data centers, manufacturing equipment and commercial kitchens that exceeds 5% of the total energy use of the whole building)
- Miscellaneous loads

604.4 Energy type metering. Buildings shall be provided with the capability to determine energy use and peak demand for each of the following energy types. Utility energy meters shall be permitted to be used to collect whole building data, and shall be equipped with a local data port connected to a data acquisition system in accordance with Section 604.5.

- Gaseous fuels
- Liquid fuels
- Solid fuels
- Electric power
- District heating and cooling
- Combined heat and power
- Renewable and waste energy

604.5 Energy load type sub-metering. Energy sub-metering of energy loads as defined in Section 604.3 (above) shall be provided for buildings greater than 25,000 square feet in total building floor area. In buildings with tenants, the metering shall be collected for the entire building and for each tenant individually. Tenants shall have access to all data collected for their space.

- For buildings that are less than 25,000 square feet in total building floor area, the energy distribution system shall be designed and constructed in such a way as to accommodate the future installation of sub-meters.

604.6 Minimum energy measurement and verification. All required meters and sub-meters shall be connected to a data acquisition and management system capable of storing not less than 36 months worth of data collected by all *meters* and transferring the data in real time to a display as required in Section 604.7.

604.7 Energy display. A permanent, readily accessible and visible energy display shall be provided adjacent to the main building entrance or on a publically available internet website that is capable of showing the current energy demand for the whole building for each fuel type and the total energy usage for the previous 18 months.

Section 605: Automated Demand Response Infrastructure (*Prescriptive and Performance Compliance Path*)

A *building energy management and control system* (*EMCS*) shall be provided and integrated with *building* HVAC systems controls and lighting systems controls to receive an open and interoperable *automated demand response* (*Auto-DR*) relay or internet signal, with exceptions. (Section 605.1)

605.1 Establishing an open and interoperable automated demand response (Auto-DR) infrastructure. When an automated demand response (Auto-DR) program is offered by the local electric utility provider, a building energy management and control system (EMCS) shall be installed and integrated with building HVAC systems controls and lighting systems controls to receive an open and interoperable Auto-DR relay or internet signal. Building HVAC and lighting systems and specific building energy-using components shall incorporate preprogrammed demand response strategies that are automated with a demand response automation internet software client.

605.3 Heating ventilating and air-conditioning (HVAC) systems. The Auto-DR strategy for HVAC systems shall be capable of reducing the building peak cooling or heating HVAC demand by not less than 10 percent when signaled from the electric utility through any combination of specified strategies and systemic adjustments specifed in the code.

605.4 Lighting. The *Auto-DR* system shall be capable of reducing total connected power of lighting in Group B office spaces by not less than 15 percent, with following exceptions.

Table 8-8 *(Continued)*

- Buildings or portions associated with lifeline services
- Luminaires on emergency circuits
- Luminaires located in emergency and life safety areas of a building
- Lighting in buildings that are less than 5,000 sq. ft. in total area
- Luminaires located within a daylight zone that are dimmable and connected to automatic daylight controls
- Signage used for emergency, life safety or traffic control purposes

Section 606: Building Envelope Systems (*Prescriptive Compliance Path*)

606.1.1 Insulation and fenestration criteria.

606.1.2 Air Leakage.

606.1.2.5 Vestibules. Doors that separate conditioned spaces from the exterior shall be protected with an enclosed vestibule, with all doors opening into and out of the vestibule equipped with self-closing devices.

Section 607: Building Mechanical Systems (*Prescriptive Compliance Path*)

607.2 HVAC equipment performance requirements.

607.3 Ventilation

607.4 Duct and plenum insulation, sealing and testing.

607.5 HVAC piping insulation.

607.6 Economizers.

607.7 Variable air volume (VAV) fan control.

Section 608: Building Service Water Heating (*Prescriptive Compliance Path*)

608.2 Service water heating equipment performance.

608.3 Pools, hot tubs, and spas.

608.4 Snowmelt systems.

608.5 Rough-ins for future solar water pre-heat.

608.6 Waste water energy recovery system.

608.7 Service water heating piping insulation.

608.8 Circulating hot water systems. Circulating hot water systems shall be provided with an automatic or readily accessible manual switch to turn off the hot water circulation pump when not in use. Controls that allow a continuous timer, or water temperature-initiated operation of a circulating pump are prohibited. Gravity or thermosyphon circulation loops are prohibited. Pumps on circulating hot water systems shall be activated on demand by either a hard-wired or wireless activation control.

Section 609: Building Electrical Power and Lighting Systems (*Prescriptive and Performance Compliance Path*)

609.2 Sleeping units in hotels, motels, boarding houses or similar buildings shall have a captive key control at the main room entry that controls all permanently wired luminaires and switched receptacles, except those in the bathroom(s). Bathrooms within sleeping units shall be equipped with occupant sensors that control all permanent wired luminaires.

609.3 Occupant sensor controls shall be provided to automatically reduce connected lighting power by at least 50 percent during periods when occupants are not present in all of the following locations.

- Corridors and enclosed stairwells
- Storage and stack areas not open to the public
- Parking garages

609.4.1 Exterior lighting shall be controlled by a time switch and configured so that the total exterior lighting power is automatically reduced by at least 30 percent within two hours after facility operations conclude. Exceptions include lighting which is connected to occupant sensor controls, lighting within means of egress, and solar powered luminaires.

609.4.2 The lighting of building facades, signage, and landscape features shall be controlled by a time switch and configured so that it automatically shuts off within one hour after facility operations conclude or as established by the jurisdiction.

(Continues)

Table 8-8 (*Continued*)

609.5 Automatic daylight controls and minimum fenestration shall be provided in all daylight zones as specified in Table 609.5 of the IGCC for Sky Type A. Exceptions include spaces within dwelling or sleeping units, spaces where medical care is directly provided, and spaces with less than 90 watts of lighting installed in the daylight zone.

609.6 Receptacles and electrical outlets controlled by an occupant sensor or time switch shall be provided in the specified locations of IGCC Section 609.6.

Section 610: Specific Appliances and Equipment (*Prescriptive and Performance Compliance Path*)

610.2 Appliances and equipment not covered by the Federal Efficiency Standards (Table 610.1) that are permanently connected to the *building* energy supply system(s) shall meet the specified requirements of Sections 610.2.1 thru 610.2.4 including the following equipment.

- Elevators
- Escalators and moving walkways
- Commercial food service equipment (not less than 50% of the aggregate rate power shall be ENERGY STAR-eligible food service equipment)
- Conveyors

610.3 Portable appliances and equipment not covered by the Federal Efficiency Standards (Table 610.1) that are not permanently connected to the *building* energy supply system(s) shall be ENERGY STAR qualified for at least 50 percent of the aggregate rated power of all portable appliances and equipment in the building or tenant space. Such appliances and equipment include the following:

- Residential service appliances
- Commercial service appliances
- Consumer electronics
- Office machines and equipment

Section 611: Building Renewable Energy Systems (*Prescriptive and Performance Compliance Path*)

611.1, 611.2, 611.3, 611.4 Each *building* or surrounding lot or *building site* where there are multiple *buildings* on the *building site* shall be equipped with at least one of the following renewable energy systems.

- Solar photovoltaic system sized to provide at least two percent of the total estimated annual electric energy consumption of the *building*, or collective *buildings* on the *building site*; or cover an area equal to not less than 30 percent of the gross *building* roof area.
- Solar water heating equipment sized to provide at least ten percent of the *building's* annual estimated hot water energy usage.
- Wind energy system sized to provide at least two percent of the total estimated annual electric energy consumption of the *building*, or collective *buildings* on the *building site*; or cover the maximum available gross *building* roof area.
- Provide at least four percent of the total annual building energy consumption for renewable generation in a five-year commitment to a renewable energy credit (REC) purchase.

611.5 On-site renewable energy systems shall be *metered* and monitored in accordance with Section 604.

Section 612: Energy Systems Commissioning and Completion (*Prescriptive and Performance Compliance Path*)

612.1 Within 60 days of final mechanical inspection, the *registered design professional* shall provide evidence of mechanical systems *commissioning* and completion in accordance with the *International Energy Conservation Code* and specified provisions contained in this section.

612.3 Prior to issuance of a *certificate of occupancy*, the *registered design professional* shall provide evidence of lighting and electrical systems *commissioning* and completion in accordance with the *International Energy Conservation Code* and specified provisions contained in this section.

612.4 Prior to issuance of a *certificate of occupancy*, the *registered design professional* shall provide evidence of *building thermal envelope* systems *commissioning* and completion in accordance with the *International Energy Conservation Code* and specified provisions contained in this section.

INDOOR ENVIRONMENTAL QUALITY, HEALTH, AND WELL-BEING

THE IMPACT OF INDOOR ENVIRONMENTAL QUALITY (IEQ)

Indoor environmental quality (IEQ) is the most direct human-related issue of green buildings, as it directly affects the health, comfort, and well-being of building occupants. The average American spends 90% of his or her life in sealed and often poorly ventilated buildings where exposure to pollutants can be two to five times higher than outdoor levels.[1] IEQ covers a wide range of issues, including effective ventilation, moisture control, low-emitting materials, air filtration, pollutant source control, acoustics, views to the outdoors, daylighting, and controllability of thermal and lighting environments. Table 9-1 identifies building elements, activities, and conditions affecting IEQ.

Table 9-1
Building Elements Affecting IEQ

Outdoor Environment
Climate
Ambient air quality: Particles and gases from combustion, industrial processes, plant metabolism (pollen, fungal spores, bacteria), human activities
Soil: Dust particles, pesticides, bacteria, radon
Water: Organic chemicals including solvents, pesticides, by-products of treatment process chemical reactions
Building Envelope
Envelope: Daylight, infiltration, water intrusion
Building Contents
Materials and finishes: Emissions
Furnishings: Emissions
Equipment and appliances
Occupants and their Activities
Occupant activities: Occupational, educational, recreational, domestic
Metabolism: Activity and body characteristics dependent
Personal hygiene
Occupant health status
Operation and Maintenance
HVAC equipment: Operator training
Ventilation performance: Monitoring, verification and corrective measures
Housekeeping and cleaning

© CENGAGE LEARNING 2012

---

[1]U.S. Environmental Protection Agency, *Heathy Buildings, Healthy People: A Vision for the 21st Century* (Washington, DC: U.S. EPA, 2001).

Many materials and products used in our buildings contain chemicals that not only pollute our physical environment but also our bodies. Bacteria and natural toxins have been part of our evolutionary past for millions of years, long enough for our bodies to develop defenses. However, most synthetic substances in our buildings were introduced since World Wars I and II and have not been around long enough for our bodies to adapt physiologically.[2] These substances can enter our bodies through the air we breathe, the water we drink, and through direct skin contact, without being detected by our senses. Continuous exposure to hazardous substances can lead to learning disabilities, cancers, and illnesses caused by damage to the nervous system.[3]

IEQ also affects human performance and productivity. Increases in ventilation rates, low-emitting materials, daylighting, and better comfort controls (temperature, humidity, ventilation) can improve work performance in offices and schools. With employees' salaries making up a significant cost of running a business, it makes prudent business sense to keep staff healthy and comfortable by improving and maintaining the quality of the indoor environment. The potential savings and productivity gains from improved indoor environmental quality in the United States are estimated to be $6 to 14 billion from reduced respiratory disease, $1 to 4 billion from reduced allergies and asthma, $10 to 30 billion from reduced sick-building-syndrome symptoms, and $20 to 160 billion from direct improvements in worker performance that are unrelated to health.[4] It is the responsibility of the design and building community to design and construct buildings that support human health, wellness, and productivity.

DESIGN WITH NATURE

Chemical and Pollutant Sources

Modern chemistry has made possible many technical and medical advances over the last two centuries. Engineered chemicals are an integral part of everyday products such as cosmetics, electronics, processed foods, and building materials. Of the more than 80,000 human-made chemicals registered with the Environmental Protection Agency (EPA), about 15,000 are produced each year in major quantities, and only 43% of these have been properly tested for human toxicity.[5] Many of these chemical materials originated from the petroleum and plastics industries, weapons development, pesticide production, manufacturing, and health-care waste.[6]

---

[2]Keeler, Marian and Burke, Bill, *Fundamentals of Integrated Design for Sustainable Building* (Hoboken, NJ: Wiley, 2009).

[3]Ibid.

[4]Environmental Health, Safety and Quality Management Services for Business and Industry, and Federal, State and Local Government, IAQ Fact Sheet, March 9, 2006.

[5]Moyers, Bill and Jones, Sherry, *Trade Secrets: A Bill Moyers Report* (New York: A Production of Public Affairs Television, 2001).

[6]Keeler, Marian and Burke, Bill, *Fundamentals of Integrated Design for Sustainable Building* (Hoboken, NJ: Wiley, 2009).

By-products of the Petroleum Industry

Synthetic petroleum products were originally developed during World Wars I and II to help fuel the war machine, including chemical warfare agents. Today, petroleum is widely used in solvents, fuels, lubricants, adhesives, asphalt, synthetic fibers, plastics, paint, detergents, pharmaceuticals, and fertilizers (see Table 9-2).[7]

Oil-based products in an enclosed environment can release transitory organic compounds either as direct emissions or as a result of a chemical reaction with other materials such as concrete. Many of these pollutants irritate the mucous membranes and can produce numerous symptoms, including irritation in the eyes, nose, and throat; unusual tiredness; headache; giddiness; sickness; and respiratory illnesses. More serious emission exposure can cause allergies, cancer, or embryonic malformation (see Table 9-3).[8]

Plastics

Plastics were the miracle materials of the post-World War II years. Early developments in plastics, including polystyrene (1839), phenol formaldehyde resin (1907), and polyurethane (1937), were followed in the 1950s by the development of numerous plastics, including high-density polyethylene (HDPE), polypropylene (PP), and polyvinyl chloride (PVC). Plastic is an extremely versatile material that can be molded, extruded, and used for a wide array of applications.[9]

Plastics are divided into two categories: thermoplastics and thermosetting plastics (see Table 9-4). Thermoplastics leave the factory complete, but can be worked to a certain extent with pressure and heat, and can be cut. Common thermoplastics in the building industry are PVC, polypropylene, polyethylene, and polystyrene. Thermosetting plastics differ from thermoplastics in that they are not finished products. The product is completed by smaller companies or at the building site where hardeners are added using two component plastics, among them polyester, epoxy, and polyurethane. Synthetic rubbers are a subgroup of thermosetting plastics.[10]

Table 9-2
Buildings Materials from Oil and Gas

Material	Areas of Use
Bitumen	Vapor barrier, damp-proofing, mastic
Asphalt	Mastic, vapor barrier, damp-proofing
Organic solvents	Paint thinner, glue, mastic
Plastics	Sheeting, window frames, wallpaper, cladding, flooring, thermal insulation, electric insulation, DWV piping, door and window composition
Other chemicals	Additives in concrete and plastics, organic pigments, impregnation, additives and binders in paint and glue

© CENGAGE LEARNING 2012

[7]Keeler, Marian and Burke, Bill, *Fundamentals of Integrated Design for Sustainable Building* (Hoboken, NJ: Wiley, 2009).

[8]Berge, Bjorn, *The Ecology of Building Materials* (Burlington, MA: Architectural Press, 2000).

[9]Keeler, Marian and Burke, Bill, *Fundamentals of Integrated Design for Sustainable Building* (Hoboken, NJ: Wiley, 2009).

[10]Berge, Bjorn, *The Ecology of Building Materials* (Burlington, MA: Architectural Press, 2000).

Table 9-3
Oil-Based Chemicals with High Environmental Risk

Oil based chemical	Areas of Use	Environmental Effects
Formaldehyde	Glue in OSB and plywood	Carcinogenic; allergenic; irritates air inhalation routes; posionous to water organisms
Phenol	Glue in laminated lumber	Carcinogenic; mutagenic; poisonous to water organisms
Chloroprene	Synthetic rubber, glue	Carcinogenic; damages liver and kidneys; irritates inhalation routes
Butadiene	Synthetic rubber (SBR)	Probably carcinogenic
Vinyl chloride	Polyvinyl chloride(PVC)	Persistent carcinogenic; can cause damage to liver, lungs, skin and joints; irritates inhalation routes; poisonous to water organisms
Ethylene (ethene)	Polyethylene	Probably carcinogenic
Propylene (propene)	Polyethylene	Probably carcinogenic
Phthalates	Softeners in plastics	Presistent; irritates the mucous membranes; allergenic; probably carcinogenic; environmental oestrogen: damages reproductive organs
Amines	Silicone, polyurethane, epoxy	Irritate inhalation routes; allergenic; possibly mutagenilc; very acidifying in water
Epichlorohydrin	Epoxy	Carcinogenic; highly poisonous to water organisms
Acrylonitrile	Synthetic rubber	Carcinogenic; highly poisonous to water organisms
Acrylic acid	Acrylic plastics and paints	Poisonous to water organisms
Styrene	Polystyrene, polyester, synthetic rubber (SBR)	Irritates air inhalation routes; damages the reproductive organs
Isocyanate (TDI, MDI, etc.)	Polyurethane, glue	Strongly allergenic; difficult to break down; irritates skin and mucous membranes
Alkyl phenol toxilates	Pigment paste, alkyd varnish	Environmental oestrogen; damages reproductive organs

© CENGAGE LEARNING 2012

Plastics can be problematic in manufacturing and building applications. Many of plastic's components and additives are considered harmful and pose a threat to human health. Phthalates are plastics additives that are added to the PVC production process because they make the material more pliable. Studies conducted in 2005 determined that phthalates found in household dust caused abnormal genital and reproductive development in rats.[11] Phthalates are used in building materials, including vinyl wall coverings and various types of resilient flooring (see Table 9-5).

Another chemical found in polycarbonate plastic is bisphenol A (BPA), which leaches from the material when heated. It mimics the estrogen hormone and in

---

[11]Shanna H. Swan, et al., "Decrease in Anogenital Distance Among Male Infants with Prenatal Phthalate Exposure," *Environmental Health Perspective* 113, no. 8 (August 2005): 1056–1061. See also Julia R. Barrent, "Phthalates and Baby Boys: Potential Disruption of Human Genital Development," *Environmental Health Perspective* 113, no. 8 (August 2005): A542.

non-human test subjects, it causes harm to developing fetuses and poses potential transgenerational reproductive harm. Toxic plastics by-products such as dioxin, lead, various phthalates, and plasticizers bioaccumulate in our bodies and pose a serious threat to human health.[12] Through off-gassing and flaking, versatile products such as PVC and their by-products and additives can lead to cancer, endocrine disruption, endometriosis, neurological damage, birth defects, impaired child development, and reproductive and immune system damage.[13]

Table 9-4
Use of Plastics in Building Construction

Type of Plastic	General Areas of Use
Thermoplastics	
Polyethylene (PE):	
Hard	Drainpipes, water pipes, interior furnishings and detailing
Soft	Sheeting (vapor barrier, in foundation work, false ceilings), cable insulation
Polyisobutylene (PIB)	Roofing felt
Polypropylene (PP)	Sheeting, boards, pipes, carpets (needle-punched carpet), electric fittings, electric switches, cable insulation
Polyamide (PA)	Pipes, fibre, carpets (needle-punched carpet), electric fittings, electric switches, cable insulation; tape
Polyacetal (POM)	Pipes, boards, electric fittings
Polytetrafluorethylene (PTFE)	Thermally insulated technical equipment, electrical equipment, gaskets
Polyphenyloxide (PPO)	Thermally insulated technical equipment
Polycarbonate (PC)	Greenhouse glass, roof lights
Polymethyl methacrylates (PMMA)	Rooflights, boards, flooring, bath tubs, paint
Methyl metacrylate (AMMA)	Paint
Polyvinyl chloride (PVC)	Sheeting, boards, sections/profiles, window frames, pipes, cable, artificial leather, flooring, wallpapers, gutters, sealing strips
Polystyrene (PS, XPS, EPS)	Sheeting, thermal insulation (foamed), electrical insulation, light fittings
Acrylonitrile butadiene styrene (ABS)	Pipes, door handles, electric fittings, electrical switches
Polyvinyl acetate (PVAC)	Paint, adhesives
Ethylene vinyl acetate sampolymer (EVA)	Paint, adhesives
Cellulose acetate (CA)[1]	Tape, sheeting
Polyacryl nitrile (PAN)	Carpets, reinforcement in concrete
Thermosetting plastics:	
Butadiene styrene rubber (SBR)	Flooring, sealing strips

[12]Hitti, Miranda, *Plastic Chemical Safety Weighed,* http://www.ewg.org/node/22367 (August 8, 2007).

[13]Healthy Building Network, "PVC Facts," http://healthybuilding.net/pvc/facts.html.

Table 9-4 (Continued)

Butadiene acrylonitrile rubber (NBR)	Hoses, cables, sealants
Chloroprene rubber (CR)	Sealing strips
Ethylene propylene rubber (EPDM)	Sealing strips
Butyl rubber (IIR)	Sealing strips
Silicone rubber (SR)[2]	Electrical insulation, sealants
Polysulphide rubber(T)	Sealants
Casein plastic (CS)[3]	Door handles
Phenol formaldehyde (PF) (Bakelite)[4]	Handles, black and brown electrical fittings, thermal insulation (foamed), laminates, adhesives for plywood and chipboard
Urea formaldehyde (UF)	Light-colored and white electrical fittings, socket outlets, switches, adhesive for plywood and chipboard, toilet seats, thermal insulation (foamed)
Melamine formaldehyde (MF)	Electrical fittings, laminates, adhesives
Epoxide resins (EP)	Filler, adhesives, paint, floor finishes, clear finishes, moulding of electrical components
Polyurethane (PUR)	Thermal insulation (foamed), adhesives, clear finishes, floor finishes, moulding of electrical components, paint, sealants
Unsaturated polyester (UP) (reinforced with fibreglass)	Roof lights, window frames, gutters, adhesives, clear finishes, floor finishes, rooflight domes, tanks, bath tubs, boards, paint

© CENGAGE LEARNING 2012

Note:
(1) Based on cellulose
(2) Based on silicon dioxide, but polymerization requires the help of hydrocarbons
(3) Based on milk casein with the additional help of formaldehyde
(4) Bakelite is the trade name for phenolic materials manufactured by Bakelite Xylonite Ltd.

Table 9-5
Use of Plastics in a Typical Dwelling

Use	Weight (lbs)	Percentage
Flooring	1764	30
Glue, mastics	1543	26
Piping	937	16
Paint, filler	606	10
Wallpaper, sheeting (e.g., vapor barrier)	441	8
Thermal insulation	220	4
Electrical installation	220	4
Cover strips, skirting, etc.	110	2
Total	5841	100

© CENGAGE LEARNING 2012

Sick Building Syndrome, Building-Related Illness, and Chemical Sensitivity

Sick building syndrome (SBS) relates to a collection of widely varying symptoms incurred by building occupants, including headaches, irritated eyes, nausea, fatigue, dizziness, coughing, and throat irritation. These symptoms tend to go away or improve significantly when individuals leave the building. SBS is different from building-related illness (BRI), where permanent health impacts can be directly linked to an unsafe exposure in a building, such as the case with Legionnaire's disease.

The issue of SBS is complicated by enormous variation in how individuals react to chemicals and their by-products. Many people seem unaffected by relatively high levels of contaminants, whereas a growing percentage of the population is extremely sensitive to even tiny quantities of contaminants. This condition is referred to as multiple chemical sensitivity (MCS).

Many individuals suffering from MCS can trace their sensitivity to a particular event in which they were exposed to a high dose of a toxic compound, which greatly weakened their immune systems. Some experts believe that a low cumulative exposure to chemicals over a long period of time gradually weakens our immune systems and can result in chemical sensitivity.[14] Indeed, some suggest that the incidence of MCS in our population is increasing at an alarming rate. In this sense, today's chemically hypersensitive individuals may unknowingly be serving as canariies (early detectors) at school, and work, warning others of the eventual consequences of continued exposure to contaminated air.

Emission, Transmission, Deposition, and Exposure

Pollutants and chemicals begin their journey to our bodies through the act of being released (emission) from materials, substances, and processes. They travel (transmission) over distance and time through air, water, and soil, in the buildings we occupy (deposition), and in the food we consume.[15]

Damage from chemicals can be classified into four categories: carcinogens (causing cancer), teratogens (causing birth defects), developmental and reproductive toxicants (causing abnormal

Defining Green

Chemical Body Burden

Sometimes referred to as *total chemical load*, chemical body burden is the capacity of our bodies to absorb chemicals from the environment. When the total load is reached, our immune systems are compromised and the body becomes susceptible to environmental illnesses based on one's state of health.

---

[14]Gibson, Pamela Reed, *Multiple Chemical Sensitivity* (Oakland, CA: New Harbinger Publications, 2000).

[15]Keeler, Marian and Burke, Bill, *Fundamentals of Integrated Design for Sustainable Building* (Hoboken, NJ: Wiley, 2009).

fetal development or harm to reproductive system), and endocrine disrupters (interfering with hormone function).[16]

Chemical body burden is defined as "the presence of hazardous chemicals and their residues in humans."[17] Some substances build up in the body because they are stored in fat or bone and leave the body at a very slow rate. The amount of a particular chemical stored in the body at a particular time is a result of exposure. Body burdens can be the result of long- or short-term storage, for example, the amount of carbon monoxide in the blood or the amount of lipophilic (fat-loving) substance such as polychlorinated biphenyls (PCBs) in adipose tissue.[18] Following are key concepts related to environmental pollution and body burdens:[19]

1. Frequency of exposure: A small dose can have a minimal effect after a single exposure but a disastrous effect if this exposure recurs over an extended period of time

2. Chemical reaction with other substances: A small dose of one substance can activate or magnify the effect of another substance. The risks of many building-industry chemicals have not been evaluated, let alone the risks of combinations of substances. A combination of small doses of various substances can develop into an unexpected hazard.

3. Biomagnification: Dilution is not always the solution to pollution. Even if a finished product is benign, the toxic substances released into the environment during the manufacturing, recycling, and disposal processes must be carefully considered. A small dose can be biomagnified into an environmentally persistent poison.

4. Trace amount: Many synthetic chemicals are harmful in minute doses. Lead causes brain damage at only 21 parts per billion (ppb). A 2005 study conducted by the Centers for Disease Control and Prevention (CDC) found that the blood of many people today contains levels of toxic chemicals such as perfluorochemicals (PFCs) (which can be found in stain repellents) and Badge-40H (a plastics ingredient) at 45 ppb and greater.[20]

5. Safe levels: For many chemicals, there are no safe levels prescribed by law simply because conclusive scientific studies have not been commissioned or turned into legislation. For other chemicals, safe levels established by the EPA or Occupational Safety and Health Administration (OSHA) have been most often designed for average adult males rather than for more vulnerable

---

[16]Keeler, Marian and Burke, Bill, *Fundamentals of Integrated Design for Sustainable Building* (Hoboken, NJ: Wiley, 2009).

[17]Lloyd, Robin, "'Chemical Body Burden' Researchers and Advocates Raise Questions About Biomonitoring Studies and Hazards Regulations," *Scientific American* (February 20, 2011).

[18]Green Facts: Facts on Health and the Environment - http://www.greenfacts.org/glossary/abc/body-burden.htm (accessed 9/16/11).

[19]Keeler, Marian and Burke, Bill, *Fundamentals of Integrated Design for Sustainable Building* (Hoboken, NJ: Wiley, 2009)

[20]National Center for Environmental Health Division of Laboratory Services, *Third National Report on Human Exposure to Environmental Chemicals*, Department of Health and Human Services, NCEH Pub. No. 05-0570 (Washington, DC: Centers for Disease Control and Prevention, 2005), 1.

individuals such as children, females, elderly, or individuals who become sick with a weak immune systems.

6. We are not all the same: Some people are more susceptible than others to developing diseases. People differ widely in their ability to eliminate cancer-causing agents to which they have been exposed. One individual may weather exposure that would be fatal to another.

7. Our own resilience changes over time: A person may respond differently to exposure at different periods in his or her life. A child's developing brain can be damaged by a trace amount of lead, whereas an adult's brain may not be harmed at that same level of exposure. An older person, whose body has been compromised by environmental stresses over a lifetime, may not be able to resist harm with as mush resilience as a young adult. The health of the individual and the time of life at which an exposure occurs determines the magnitude of its effect.

8. Delayed symptoms of harm: An individual who has had a prolonged exposed to a carcinogen may not experience symptoms for many years. Most cancers, for instance, take 10 to 15 years to manifest. The more time that elapses between the exposure and the symptom, the more difficult it is to link them as cause and effect.

For 30 years, the National Center for Environmental Health at the Center for Disease Control and Prevention (CDC) has been biomonitoring human subjects to understand the extent of human exposure to toxic substances in the environment. Biomonitoring involves the periodic measurement of chemical substances in blood, urine, breast milk, hair, organs, and tissues. The following chemicals have shown up as a body burden:[21]

- PCB, a chlorinated compound used as a coolant in electrical equipment
- Dichloro-diphenyl-trichloroethane (DDT), a pesticide banned in the 1970s
- Polybrominated diphenylethers (PBDE), a fire-retardant treatment for building materials
- Dioxins (used in the manufacture of PVC and other plastics and released during their incineration)
- Phthalates (a plasticizing component of PVC and used in cosmetics)
- Triclosan (a common antibacterial agent)
- Furans, metals, organochlorine and organophosphate insecticides

The "Wingspread Consensus Statement on the Precautionary Principle" (January 1998)[22] declared the following imperative:

"When an activity raises threats of harm to human health or the environment, precautionary measures should be taken even if some cause and effect relationships are not fully established scientifically. In this context, the proponent of an activity, rather than the public, should bear the burden of proof. The process of applying the precautionary principle must be open, informed

---

[21]Moyers, Bill and Jones, Sherry, *Trade Secrets: A Bill Moyers Report* (New York: Public Affairs Television, 2001).

[22]Science and Environmental Health Network, "Wingspread Statement on the Precautionary Principle" (January 1998), http:www.sehn.org.

Table 9-6
Four Tenets of the Precautionary Principle[23]

1. People have a duty to take anticipatory action to prevent harm.

2. The burden of the proof of harmlessness of a new technology, process, activity, or chemical lies with the proponents, not the general public.

3. Before using a new technology, process, or chemical or starting a new activity, people have an obligation to examine a full range of alternatives including the alternative of not doing it.

4. Decisions applying the Precautionary Principle must be open, informed, and democratic and must include the affected parties.

and democratic and must include potentially affected parties. It must also involve an examination of the full range of alternatives, including no action."

The precautionary principle requires the exercise of caution when making decisions that may adversely affect human health and natural ecosystem cycles. The precautionary principle should be applied to new technologies and chemicals in order to minimize risks to future populations, both human and nonhuman, from the consequences of technologies and chemicals that are not fully understood (see Table 9-6).

Chemicals and Pollutants in the Construction Industry

There are hundreds of chemicals in the indoor environment. Most building products emit volatile organic compounds (VOCs), resulting in indoor concentrations that are often higher than those outdoors. Hazardous airborne pollutants can be both emitted and stored in buildings. Sources of emissions range from the flame retardants in fabrics and upholstery to PVC in shower curtains to the formaldehyde in composite wood products. Besides controlling chemicals in the indoor air, it is equally important to control microbial growth inside buildings. Water intrusion through the building envelope, condensation inside buildings, and plumbing leaks are some of the causes of mold growth. Other important indoor pollutants include particulate matter, CO_2, and ozone (see Figure 9-1 and Table 9-7).

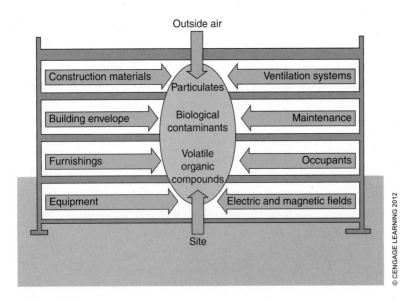

© CENGAGE LEARNING 2012

Figure 9-1

Sources of indoor pollutants.

---

[23]Kibert, Charles J., *Sustainable Construction: Green Building Design and Delivery* (Hoboken, NJ: Wiley, 2008).

Table **9-7**
Sources of Chemical and Their Health Risks[24]

Chemical	Source Products	Health Risks
Polybrominated diphenylethers (PBDEs)	Flame retardants in furniture foam and upholstery fabrics	Brain development, thyroid function, suspected carcinogen
Polychlorinated biphenyls (PCBs)	Insulation, lubricants (banned in the US)	Cancer, nervous system damage
Polychlorinated naphthalenes (PCNs)	Wood preservative, varnished, fabric dyes	Liver and kidney damage
Perfluorinated chemicals (PFCs)	Fabric and carpet stain repellents	Cancer, birth defects
Alkylphenols	Plastics, maintenance products	Endocrine disruption
Di(2-ethylhexyl) phthalate (DEHP)	Carpet, upholstery fabrics, shower curtains	Endocrine disruption, sperm damage

Most materials have some level of emissions, and these materials all contribute to some degree to the deterioration of indoor air quality. Proper selection of materials can provide for a healthier indoor environment. Additional concern is that some materials act as "sinks" for emissions for other materials or contaminants that enter the building from other sources. For example, many building materials readily absorb VOCs and re-release them into the air. The first step is to design for the use of low-emitting materials in the following building components and systems:

1. Architectural coatings
2. Caulks, sealants, and adhesives
3. Ceiling tiles
4. Composite wood products
5. Flooring materials
6. Insulation materials
7. Office furniture systems

Sound/Noise Transmission

Any sound that is undesirable becomes noise. Noise from air-conditioning equipment, air-handling systems, lighting, transformers, motors, and other sources can cause discomfort and even health problems (see Table 9-8). Noise levels can also lead to morale and productivity issues when occupants become irritated, annoyed, and distracted from work. Fluorescent light ballasts often buzz when they are not in proper working condition, and ventilation systems produce an array of distracting sounds from vibrations to humming and whistling sounds. Improperly controlled sound transmission through floors, ceilings, and walls between dwelling units or other occupancies can lead to poor indoor environmental conditions particularly between neighbors. The amount of reflective and absorptive surfaces in a room can be conducive or detrimental

---

[24]Rudel, Ruthann A., David E. Camann, John D. Spengler, Leo R. Korn, and Julia G. Brody, "Phthalates, Alkylphenols, Pesticides, Polybrominated Diphenyl Ethers, and other Endocrine-Disrupting Compounds in Indoor Air and Dust," *Environmental Science & Technology* 37, part 20: 4543–4553.

Table 9-8
Illnesses That Can Be Attributed to Noise Pollution[25]

- Neuropsychological disturbances
- Headaches
- Fatigue
- Stress
- Insomnia and disturbed sleep patterns
- Mood effects, irritability, and neurosis
- Cardiac disease and cardiovascular system disturbances
- Hypertension and hypotension
- Digestive disorders, ulcers, colitis
- Endocrine and biochemical disorders
- Higher breathing and cardiac rates
- Hearing loss
- Vision modification
- Cognitive problems (including slower learning in children), social behavior disturbances

to various types of activities or tasks and can reduce or enhance speech intelligibility, which can be an advantage or a disadvantage, depending on the specific scenario.

Acoustics and sound control have long been important architectural design considerations. The IBC requires sound insulation for common interior walls, partitions, and floor/ceiling assemblies between adjacent dwelling units and between dwelling units and adjacent public areas, including corridors, stairs, elevators, and service areas. Sound transmission is rated by a sound transmission class (STC) determined by testing sound materials and assemblies in accordance with ASTM Standards E90 and E492. Noise is ubiquitous in both public and private areas and is increasingly viewed as a major public health and comfort concern. IEQ for green and sustainable buildings extends beyond the minimum measures of baseline codes by providing sound insulation measures between the exterior environments and the indoors. Sources of noise include proximity to airports, railways, major roadways, and outdoor equipment, as well as the noise of bustling restaurants, concert halls, manufacturing, and car alarms.

Light Quality

Natural sunlight is the best light source for the eyes. The ideal healthy indoor light environment is one that allows natural light indoors. Natural sunlight has an equal spectral distribution of the visible light frequencies combined to appear as white light (see Figure 9-2). In contrast, artificial light sources are limited in the frequencies

---

[25]Green Building Briefing Paper, Leonardo-Energy, January 2007, "Green Buildings: What Is the Impact of Construction with High Environmental Quality?"

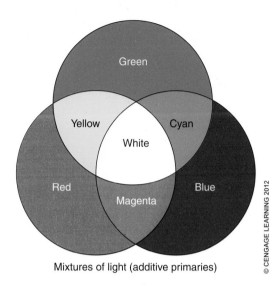

Yellow

Green

White

Cyan

Red

Magenta

Blue

Mixtures of light (additive primaries)

© CENGAGE LEARNING 2012

Figure 9-2

Visible light frequencies.

of visible light that they emit (see Table 9-9). Although incandescent lights emits a color spectrum dominated by reds and oranges, halogen lamps give the best color rendition of natural light. Fluorescent lights emit white light dominated in the blue spectrum. Fluorescents can be made to offer more in the warm color range, but the color is not natural and tends to produce a bright, sterile atmosphere.

Fluorescent lighting in office settings can often be irritating to occupants. The frequent complaints caused by too much fluorescent light are sore eyes and headache, lowered morale, and decreased productivity. Poor lighting can also affect a person's mood.[26] With the exception of glare, the eye is most comfortable with natural sunlight, which changes in intensity and color throughout the day. Because indoor artificial light is basically unchanging in color and intensity, there may be adverse effects on the health and well-being of those subjected to it on a regular and long term basis. Flickering lights can cause irritation and health problems. Ballasted lights, such as fluorescent, are prone to flickering when the ballast malfunctions. This can easily lead to sore eyes and headaches. In addition, depending on the intensity of the light, glare can lead to discomfort and headaches, especially when reading, typing, or working on a computer.

Connection to the Outdoors

Biophilia is the human need to be connected to nature and living things. Proximity to natural settings and visual access to the sky are comforting and soothing to well-being. Studies have shown that even minimal connection with nature, such as looking outdoors through a window, increases productivity and health in the workplace, promotes healing of patients in hospitals, and reduces the frequency of sickness in prisons.[27]

Table 9-9
Color Characteristics of Artificial Lighting Systems

Type of Light	Color Characteristics
Incandescent (argon-surrounded filament)	White with yellow tint
Incandescent (halogen-surrounded filament)	White
Fluorescent	White with blue tint
Mercury vapor	White with blue tint
Metal halide	White with blue-green tint
Sodium vapor (high pressure)	Amber white
Sodium vapor (low pressure)	Yellow

© CENGAGE LEARNING 2012

[26]Kibert, Charles J., *Sustainable Construction: Green Building Design and Delivery* (Hoboken, NJ: Wiley, 2008).

[27]Kahn, Peter H., "Development Psychology and the Biophilia Hypothesis: Children's Affiliation with Nature," *Developmental Review*, 17, 1997, pp. 1–61.

Biophilia

Human affinity for nature, such as natural forms, vegetation, and water features. Access to daylight and views to greenery promote stress reduction, human health, productivity, and well-being. There is an innate human need to affiliate with life and lifelike processes.

Human health, productivity, and well-being are promoted by access to natural light and views of vegetation. Connecting humans to nature is a stress reliever. The ability of humans to interact with nature, even at a distance, from inside a building, is an important consideration in design of green buildings.

Daylighting

Certain tasks benefit from daylight, whereas others do not. Learning and visually oriented tasks, such as reading, fine needlework, and drawing, benefit from daylight. Other activities, such as a surgery room, would be hampered by daylight. Good daylighting strategies involve architectural building-site design decisions. The building footprint should be placed along the east-west axis with sources of daylight from more than one side and a shallow floor plate to allow daylight to penetrate the interior (see Figure 9-3). Window dimensions should be sized in proportion to the depth of the room.

A variety of strategies can be used to enhance, diffuse, and control the effects of light. Designing sawtooth skylights that reflect light from a rooftop, or admitting light through a clerestory or monitor, are two building-design strategies that can admit light without admitting direct solar heat gain (see Figure 9-4).

Using light shelves and admitting light from higher in the room will project light deeper into a space (see Figures 9-5 and 9-6). Transoms connecting rooms with access to daylight can direct light into adjoining spaces. Consider not only how the daylight enters through the window but also how direct sunlight can be externally blocked to reduce heat gain during the summer months. Shading devices can be designed into the building envelope. Exterior louvers can provide dynamic control by tracking sun angles based on time of day and season (see Figure 9-7).

It is important to combine daylight sensors with electric lighting controls (see Figure 9-8). Lighting controls are important for saving energy as well as providing quality light. Dimming technologies can be used that range from simple manual dimming, to automatic stepped dimming, ranging from 100 to 50%, to continuous dimming by using daylight sensors or a preset schedule. Other techniques include automatic shutoffs, or more sophisticated shutoffs, with photocell sensors (see

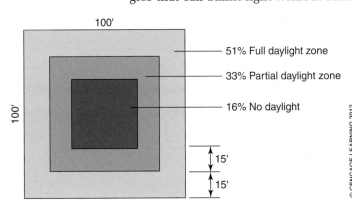

Figure 9-3

This office plan demonstrates that daylight (sidelight) penetration is foremost based on the depth of floor plate.

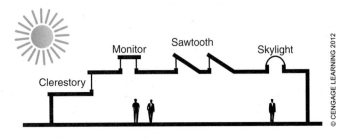

Figure 9-4

Daylight overhead strategies.

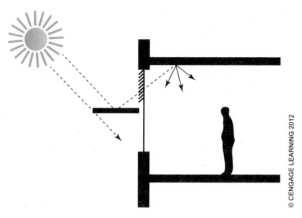

Figure 9-5

Light shelves increase the depth of the daylighting zone while providing shade and reducing glare.

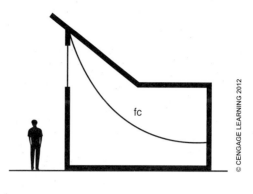

Figure 9-6

Daylight penetration and footcandle level increases with window height. Daylight glazing is most effective above 7'–6".

Figure 9-7

Exterior louvers can reduce heat gain and glare.

Table 9-10). Lighting controls cover the spectrum from simple toggle switches to sophisticated energy management systems that detect who's in a space, measure how much natural daylight is available, judge the lighting requirements based on programmed inputs, account for electricity prices and electric demand charges, and then regulate light output accordingly. Lighting design and controls are based on building function, occupant density, and visual tasks performed.

MATERIAL EMISSIONS

The selection and installation of low-emitting building materials, is the most cost-effective method of reducing the indoor contaminants when compared

Figure 9-8

Daylighting and daylight sensor controls for electric lighting.

COURTESY OF ANTHONY FLOYD

Table 9-10
Lighting Control Strategies

Strategy	Description and Explanation of Energy Savings	Potential Savings in Lighting Energy Consumption
Occupancy control	Turns lights on when needed and off when not needed	Depends on occupancy patterns, occupant energy awareness, etc.
Manual	User operates switches	10–50% (compared to no switching)
Automatic	Switching automatic based on occupancy sensors (usually automatic on-off or manual on/auto off, sometimes stepped or dimming control)	Up to 80% (compared to manual switching)
Timing control	Turns lights on and off at predetermined times, or automatically switches lights off after a time delay	Depends on occupancy patterns; typically 10–50% (compared to manual switching)
Manual dimming control	Allows user to set the light level at less than full brightness	Depends on daylight availability, personal preference, occupant energy awareness, etc.
Stepped	Reduces light level by switching off banks of lights or some lamps in fixtures	10–50%
Continuous dimming	Reduces power to all lights in a smooth continuum (requires dimming ballasts for ballasted lamps)	10–50%
Daylighting control	Turns off or dims lights when adequate daylight is present	Depends on how well daylight illuminates space
On/Off	Turns off all lights (common with outdoor lighting)	Up to 50% (compared to manual switching)
Stepped	Turns off banks of fixtures or certain lamps within fixtures	Up to 70% (compared to manual switching)

Table 9-10 (*Continued*)

Continuous dimming	Reduces power to all lights in a smooth continuum (requires dimming ballasts)	Up to 80% under optimum conditions of good daylighting and typical work hours
Lumen daintenance control	Light output dimmed up to 40% after relamping when lumen output higher than needed, then gradually increased to full power as lamps age and lumen output drops	10–20% (with higher-quality lamps, such as T-5 high output fluorescents, lumen depreciation is low, so savings will be minimal)
Tuning control	Light levels adjusted to match occupants needs or desire; can be done by initial calibration or directly by users	10–50% depending on how significantly space is overlit and how well managed by users
Adaptation compensation control	Human eyes have tremendous ability to adapt to light intensity. When it's bright outside, higher light levels may be required indoors, depending on tasks and proximity to windows. Conversely, at night you have a lower physiological expectation for less light. This lighting control strategy involves reducing light levels at night.	Saves nighttime energy in facilities operating at night
Stepped	Turns off banks of lights or certain lamps within fixtures	10–40%
Continuous dimming	Reduces power to all lights in a smooth continuum (requires dimming ballasts)	10–40%
Load shedding	Automatically dims lights or turns off unneeded lights during peak electrical demand periods	Saves minimal electricity but can reduce demand charges significantly

© CENGAGE LEARNING 2012

to increased ventilation or possible material replacement after occupancy.

There are a number of emission-testing and product-certification programs, which vary in their scope and approach. Caution should be exercised when interpreting or comparing program results. Laboratory testing conditions, target compounds, and their concentrations can vary from program to program.[28] In most certifications programs, only a pass-fail report is issued, and the emissions-test data and compound concentrations are not disclosed. Efforts to harmonize these programs have been difficult due to strong proprietary processes and commercial interests.

Products should be evaluated on individual VOCs and not on the total volatile organic compounds (TVOCs). TVOC is a poor indicator of potential health and odor effects.[29] Emissions

Defining Green

Volatile organic compounds (VOCs)

Carbon-based substances that occur as gases under typical ambient air temperature and pressure. A wide range of substances are considered VOCs, including microbial VOCs, which are generated and released as a result of microbial growth.

[28]Alevantis, Leon, "Indoor Air Quality Technologies: Green Design for Long-Term Occupant Health," Chapter 7, *Fundamentals of Integrated Design for Sustainable Building* (Hoboken, NJ: Wiley, 2009).

[29]Ibid.

Table 9-11
Common Building Materials, Emissions, and Testing Programs[30]

Building Material Category	Examples	Emissions-Related Considerations	Selected Testing Programs in the United States		
			Green Guard*	Green Seal†	Other
Architectural coatings	Sealants, primers, paints, enamels, lacquers, varnishes, stain	• Use water-based products where possible • Zero or low VOC paints may contain numerous "reactive" VOCs exempt from under the Clean Air Act • Emissions depend on substrate • Consider durability • Consider sequence of construction	X	GS-11	
Caulks, sealants, and adhesives		• Use water-based or low-VOC products • Large variations in reported emissions due to type of substrate	X	Aerosol Adhesives: GS-36	
Ceiling tiles		• Formaldehyde and SVOC emissions • Large surface area of installed panels may result in: (a) large indoor concentrations of an otherwise low-emitting product and (b) sink effects	X		
Composite wood products	Particleboards, oriented strand board, fiberboard	• Specify urea formaldehyde-free products or encapsulate surfaces	X		Composite Panel Assoc: Environmentally preferable product Spec CPA 2-06 (not very stringent formaldehyde requirements)

[30]Keeler, Marian and Burke, Bill, *Fundamentals of Integrated Design for Sustainable Building* (Hoboken, NJ: Wiley, 2009).

Table 9-11 (*Continued*)

Flooring materials	Resilient tile: Laminate, hardwood, linoleum, carpet	• Consider emissions from adhesives • Test as assembly: substrates are very important • Use low-emitting cleaning products	X	• CRI Green Label Plus • Resilient Floor Covering Institute, FloorScore Program • CA Gold Sustainable Carpet Std.
Insulation materials		• Specify urea formaldehyde-free based binder in fiberglass and mineral fiber	X	
Office furniture systems	Workstations	• Specify low-emitting products • Airing out in lieu of low-emitting products is not very practical and marginally effective for some chemicals • Install office furniture systems last followed by building flush-out • Use low-emitting cleaning products		• BIFMA • CA Modular Office Furniture Specification

*Based on emissions testing. CHPS includes additional health-based criteria

†Based on total VOC concerns and reactive VOCs

from some products are initially high, followed by rapid decline such as the case with most paints. Emissions from other products may initially be moderate but decline very slowly over an extended period of time such as the case with engineered wood products with formaldehyde binders. In the selection of low-emitting materials, it is important to consider exposure, durability, and maintenance requirements. Table 9-11 lists the common building materials and related emission considerations.

Environmental Labeling, Certification, and Verification Programs

Product labeling, certification, and verification programs test or otherwise verify that a product meets specified standards or guidelines. The credibility of a certification varies depending on the relationship of the verifier to the product. First-, second-, and third-party certifications define the degree of separation between

Emission rate

The rate at which a given compound or group of compounds off-gases from a product or test sample.

The emission rate is usually measured in micrograms of compound per square meter of product surface area per hour (μg/m2•hr). It generally decreases over time, so it is often shown as a curve on a graph.

Semi-volatile organic compound (SVOC)

SVOCs don't become gaseous as readily as VOCs but are still found in indoor air. Those most commonly identified as chemicals of concern are pesticides, flame retardants, and phthalates.

the certifying agency and the company whose product is being certified. Most marketing claims and product specifications are first-party declarations that have not been independently tested or verified.[31] Second-party certifications provide more credible information by involving a industry association or contracted company in establishing qualifying criteria and verifying compliance. Second-party certifications are usually a conflict of interest because the certifying party has a direct or indirect financial interest in the outcome of the certification. A third-party certification program conducts independent product testing and certification. It avoids a conflict of interest by providing an objective and transparent process with an independently developed consensus-based set of standards.

The International Organization for Standardization (ISO) has set various standards covering three types of environmental labeling and declaration (certification) programs (see Table 9-12).

Type I labels are multi-criteria-based, third-party-certified environmental labeling programs in compliance with ISO 14024. Type II labels are single-attribute environmental claims made about products by the producer. They are not independently verified, do not use predetermined criteria, and are the least informative of the three types of environmental labels.

Type III environmental declarations present quantified environmental information on the life cycle of a product to enable comparisons between products fulfilling the same function. Such declarations are: (1) based on independently verified life-cycle assessment (LCA) data or life-cycle inventory analysis (LCI) data; (2) developed using predetermined parameters; and (3) subject to the administration of a program operator industry.

There are many national green labeling, certification, and evaluation programs that validate green product claims for low emissions of VOCs and other harmful substances (see Table 9-13). Some of the key issues associated with emissions certification are product emissions versus product content, short-term versus long-term emissions, and emerging health concerns.[31] Wet-applied products, such as paints, coatings, adhesives, and sealants, are tested for the quantity of total organic compounds (TVOCs) as opposed to measuring the

Table 9-12
Types of Environmental Labeling and Declarations

ISO Standard	Description
ISO 14024 – Type I	Multi-attribute label developed by a third party
ISO 14021 – Type II	Single-attribute label developed by the producer
ISO 14025 – Type III	Full life-cycle assessment

[31]Atlee, Jennifer and Altes, Tristan Korthals, "Behind the Logos: Understanding Green Product Certifications," *Environmental Building News*, Vol. 17, No. 1, 1/2008.

Table 9-13
Green Labeling and Certification Programs

Green Product Labels & Certifications

To download the latest version of this chart, visit us online at www.pa-greenbuildingproducts.org.

Category	Certification Information / General Information	Outside Audit	General Product Category	Created By	Number of Products Certified	Cost to Manufacturer	Certification Process (General)	ANSI	ASTM	CHPS	ISO	LEED	NSF	Other
ROOFING	**Cool Roof Rating Council** — Attribute: Roof Coolness; System: Rating; Acronym: CRRC; www.coolroofs.org		Roofing	Cool Roof Rating Council	1,300+	$1,000 to $6,000	✓ Third-Party Certification; ✓ Public Access to Method; ✓ Public Comment; Multiple Levels: Score from 0 to 1					SS		
ENERGY	**Energy Star** — Attribute: Energy; System: Label; www.energystar.gov		Appliances, Building products, Commercial food products, Home electronics, Office equipment, & Other commercial products	U.S. Environmental Protection Agency	44,000+	No cost	✓ Third-Party Certification; ✓ Public Access to Method; ✓ Public Comment; Multiple Levels: None	✓				SS & EA		
INDOOR AIR QUALITY	**FloorScore** — Attribute: Indoor Air Quality; System: Label; www.scscertified.com/gbc/floorscore.php	✓	Ceramic flooring, Laminate flooring, Linoleum, Rubber flooring, Vinyl, Wall base, & Wood flooring	Resilient Floor Covering Institute	44+	Not available	✓ Third-Party Certification; ✓ Public Access to Method; ✗ Public Comment; Multiple Levels: None		C1371, C1549, E903, & E1918	✓		IEQ		
INDOOR AIR QUALITY	**GREENGUARD** — Attribute: Indoor Air Quality; System: Label; www.greenguard.org	✓	Bedding, Building products, Cleaning & maintenance products & systems, Electronic equipment, Furniture, Office equipment, & Visual display products	GREENGUARD Environmental Institute	200,000+	$10,000 to $40,000	✓ Third-Party Certification; ✓ Public Access to Method; ✗ Public Comment; Multiple Levels: Children & Schools: Indoor Air Quality	✓	✓	Children & Schools		IEQ	140 & 332	
INDOOR AIR QUALITY	**Green Label** — Attribute: Indoor Air Quality; System: Label; www.carpet-rug.org		Carpet	Carpet and Rug Institute	Not available	Not available	✗ Third-Party Certification; ✓ Public Access to Method; ✓ Public Comment; Multiple Levels: Label & Label Plus				✓	IEQ		
INDOOR AIR QUALITY	**Indoor Advantage™** — Attribute: Indoor Air Quality; System: Label; www.scscertified.com/gbc/indooradvantage.php		Furniture & seating systems & components	Scientific Certification Systems	290+	Not available	✓ Third-Party Certification; ✗ Public Access to Method; ✓ Public Comment; Multiple Levels: None	BIFMA X7.1				IEQ		
INDOOR AIR QUALITY	**Indoor Advantage™ Gold** — Attribute: Indoor Air Quality; System: Label; www.scscertified.com/gbc/indooradvantagegold.php		Adhesives & sealants, Composite wood & agri-fiber products, Doors, Flooring underlayment, Furniture & seating, Paints & coatings, Textiles & fibers, & Windows	Scientific Certification Systems	490+	Not available	✓ Third-Party Certification; ✗ Public Access to Method; ✓ Public Comment; Multiple Levels: None				✓	IEQ		
RECYCLED CONTENT	**Recycled Material Content** — Attribute: Recycled Content; System: Label; www.scscertified.com/gbc/material_content.php		Building products	Scientific Certification Systems	350+	Not available	✗ Third-Party Certification; ✓ Public Access to Method; ✓ Public Comment; Multiple Levels: Percentage					MR		
WATER	**WaterSense** — Attribute: Water; System: Label; www.epa.gov/watersense	✓	Residential & commercial plumbing products	U.S. Environmental Protection Agency	1,000+	Determined by EPA licensed certifying bodies	✓ Third-Party Certification; ✓ Public Access to Method; ✓ Public Comment; Multiple Levels: None	✓			IEC Guide 65	WE		
VARIES	**Sustainable Attributes Verification and Evaluation™** — Attribute: Varies; System: Claim Verification; Acronym: SAVE; www.icc-es.org/save/		All products	International Code Council	2+	$5,000	✓ Third-Party Certification; ✗ Public Access to Method; ✓ Public Comment; Multiple Levels: None					Varies		
CARPET	**California Gold** — Attributes: Multiple; System: Certification; www.green-ca.gov/299/standards.htm	✓	Carpet	California Department of General Services	18+	Not available	✓ Third-Party Certification; ✓ Public Access to Method; ✓ Public Comment; Multiple Levels: Gold & Platinum					ID	140	
CARPET	**Sustainable Choice™** — Attributes: Multiple; System: Certification; www.scscertified.com/gbc/sustainable_choice.php		Carpet	Scientific Certification Systems	110+	$15,000 to $25,000	✓ Third-Party Certification; ⊙ Public Access to Method; ✓ Public Comment; Multiple Levels: Silver, Gold & Platinum						140	

SINGLE ATTRIBUTE CONSIDERED

(Continues)

Table 9-13 (Continued)

ELECTRONICS

Label	Products	Organization	Number	Cost	Certification Details
Electronic Product Environmental Assessment Tool — Attributes: Multiple; System: Certification; Acronym: EPEAT; www.epeat.net	Desktop & laptop computers, Integrated systems, & Monitors	Green Electronics Council	1,000+	$1,500 to $90,000 Annually	✓ Third-Party Certification; ✓ Public Access to Method; ✓ Public Comment; Multiple Levels: Bronze, Silver & Gold

WOOD

Label	Products	Organization	Number	Cost	Standards/Credits	Certification Details
Forest Stewardship Council Certification — Attributes: Multiple; System: Certification; Acronym: FSC; www.fscus.org	Wood, Paper, & Other wood products	Forest Stewardship Council	13,000+ certificate holders	$2,000 to $4,000 per location per year	MR	✓ Third-Party Certification; ✓ Public Access to Method; ✓ Public Comment; Multiple Levels: Forest Management, Chain of Custody, & Controlled Wood
Sustainable Forestry Initiative — Attributes: Multiple; System: Certification; Acronym: SFI; www.sfiprogram.org	Wood, Paper, & Other wood products	Sustainable Forestry Initiative	700+ certified locations	Not available		⊘ Third-Party Certification; ✓ Public Access to Method; ✓ Public Comment; Multiple Levels: Forest Certification, Chain of Custody, & Fiber Sourcing
CSA Sustainable Forest Management — Attributes: Multiple; System: Certification; Acronym: CSA/FSM; www.csa-international.org/product_areas/forest_products_marking	Wood	Canadian Standards Association (CSA) International	79.3 million hectares (out of 134.1 million hectares of Canadian forest)	Not available	D5116 & D6670	✓ Third-Party Certification; ✓ Public Access to Method; ✗ Public Comment; Multiple Levels: Chain of Custody

PRODUCT TYPE VARIES

Label	Products	Organization	Number	Cost	Standards/Credits	Certification Details
Cradle to Cradle — Attributes: Multiple; System: Certification; Acronym: C2C; www.mbdc.com/c2c	Building products	McDonough Braungart Design Chemistry	250+	$5,000+	ID	⊘ Third-Party Certification; ✓ Public Access to Method; ✓ Public Comment; Multiple Levels: Approved Ingredients, Basic, Silver, Gold, & Platinum
EcoLogo — Attributes: Multiple; System: Certification; www.ecologo.org/en	Covers product sectors: Automotive, Building & construction, Business, Cleaning & janitorial, Consumer, Electricity, Fuels & lubricants, Office furniture, Packaging & containers, & Paper; & Services in the Automotive and Printing sectors	TerraChoice Environmental Marketing Inc.	7,000+	$1,500 to $5,000	14024	✓ Third-Party Certification; ✓ Public Access to Method; ✓ Public Comment; Multiple Levels: None
Eco Options — Attributes: Multiple; System: Label; www.ecooptions.com/programs/Home_Depot	Commercial & residential building & home products	Home Depot	3,700+	$1,500		✗ Third-Party Certification; ✗ Public Access to Method; ✗ Public Comment; Multiple Levels: None

MULTIPLE ATTRIBUTES CONSIDERED

Label	Products	Organization	Number	Cost	Standards/Credits	Certification Details
Environmentally Preferable Products — Attributes: Multiple; System: Certification; Acronym: EPP; www.scscertified.com/fsc/epp_products.php	Adhesives & sealants, Cabinetry & casework, Carpet, Composite panel products, Doors, Electric power generation, Flooring, Furniture, Paints, Renewable energy, Wall coverings, & Wood treatment technologies	Scientific Certification Systems	16+	Not available	14024 & 14044	✗ Third-Party Certification; ✗ Public Access to Method; ✗ Public Comment; Multiple Levels: Yes. Inherent to LCA basis, EPP considers a variety of impacts.
Green Seal — Attributes: Multiple; System: Certification; www.greenseal.org	Building products, Cleaners, Fleet vehicle maintenance, Lodging properties, & Paper & products	Green Seal	2,500+	$3,000 to $10,000	14020 & 14024 / IEQ	✓ Third-Party Certification; ✓ Public Access to Method; ✓ Public Comment; Multiple Levels: None
National Sanitation Foundation — Attributes: Multiple; System: Certification; Acronym: NSF; www.nsf.com	Carpet, Furniture, & Resilient flooring	National Sanitation Foundation	220+	Not available	BIFMA e3 / ID (for Carpet) / 140 & 332	✓ Third-Party Certification; ✓ Public Access to Method; ✓ Public Comment; Multiple Levels: Conformant, Silver, Gold, & Platinum
SMaRT Consensus Sustainable Product Standards — Attributes: Multiple; System: Certification; Acronym: SMaRT; mts.sustainableproducts.com/SMaRT_product_standard.html	All products except airplanes & vehicles	The Institute for Market Transformation to Sustainability	16+	$7,500 to $10,000	14040 / ID	✓ Third-Party Certification; ✓ Public Access to Method; ✓ Public Comment; Multiple Levels: Sustainable, Silver, Gold, & Platinum

LEGEND: ✓ Yes | ✗ No | ⊘ Can Comply with Third Party Certification, but Not Required | ✦ Partial Via Standards | *Others Listed Here

Green Building Alliance   333 East Carson Street, Suite 331 Pittsburgh, PA 15219   Phone: 412-431-0709   Fax: 412-431-1432

Disclaimer: Green Building Alliance does not endorse any of the green building product labeling systems, certifications, or organizations listed here. This information is not a comprehensive representation of any of the included resources and only contains information collected through April 2009.

VOCs emitted from them once installed. Wet-applied products release most of their VOCs in the first few days after application. Some standards address both VOC content and VOC emissions testing to capture short- and long-term emission outcomes.[32] Chemicals and substances also interact with each other in often unpredictable ways. Paint in a container will emit one set of compounds but may react with the paper or additives in the wallboard to emit entirely different substances and produce undesirable outcomes as it cures over time.[33]

Defining Green

Threshold limit value (TLV)

Guidelines for allowable workplace exposure to a list of chemicals, published annually by the American Conference of Governmental and Industrial Hygienists (ACGIH).

Greenguard Environmental Institute (GEI)

The GEI certifies products and materials for low chemical emissions and provides a resource for choosing healthier products and materials for indoor environments. All certified products must meet stringent chemical emissions standards based on established criteria from public health agencies. The standards are primarily for building materials, finishes, interior furnishings, furniture, cleaning products, and electronic equipment. Greenguard certification is broadly recognized and accepted by building codes and green building programs worldwide.[34]

Green Seal and EcoLogo

Green Seal and EcoLogo are the only two international ecolabels in North America and are ISO 14024 Type I programs.[35] These programs follow the internationally established ISO process for open, consensus-based standard development. They consider impacts over the entire life cycle of a product and develop criteria relating to the most significant impacts. Both Green Seal and EcoLogo[36] have numerous standards with requirements set so that roughly 20% of existing products within those product groups can meet them. Both organizations also provide third-party certification for products meeting these standards. These labels tend to address

---

[32]Atlee, Jennifer, *Green Building Product Certification* (Brattleboro, VT: Building Green, Inc. 2011).

[33]Ibid.

[34]As of this writing, UL Environment, a division of Underwriters Laboratories (UL), has acquired the Greenguard indoor air quality certifications and EcoLogo. UL Environment has interest in expanding multiple multi-attribute environmental standards for specific building product categories, including drywall, insulation, and roofing. UL's history and brand recognition could help normalize the green products certification arena.

[35]ISO Standard 14024, "Environmental Labels and Declarations: Environmental Labeling Type I, Guiding Principles and Procedures," provides guidance on developing programs that verify the environmental attributes of a product through a seal of approval.

[36]As of this writing, UL Environment, a division of Underwriters Laboratories (UL), has acquired the Greenguard indoor air quality certifications and EcoLogo. UL Environment has interest in expanding multiple multi-attribute environmental standards for specific building product categories, including drywall, insulation, and roofing. UL's history and brand recognition could help normalize the green products certification arena.

a narrower range of impacts but everything that they address is mandatory, in contrast with point-based certification programs, which may address more impact areas but with fewer prerequisites.

Cradle to Cradle (C2C)

The C2C is a multi-attribute certification and label that evaluates products based on the "cradle-to-cradle" manufacturing approach developed by McDonough Braungart Design Chemistry (MBDC). Although its environmental coverage is broad in scope, C2C primarily addresses toxicity issues throughout the supply chain with an emphasis on closed-loop systems. C2C is the first certification program in the United States to address chemistry by looking at the chemical properties of products. This approach distinguishes itself from the more common method of avoiding specific chemicals, which often has the unintended effect of sending manufacturers toward substitutes that may turn out to be just as bad.

Sustainable Materials Rating Technology (SMaRT)

The SMaRT Sustainable Building Product Standard (MTS 2006:4) is a multi-attribute standard developed by the Institute for Market Transformation to Sustainability (MTS). Although the standard covers a wide range of impacts, the requirements of individual credits and overall scoring are somewhat problematic. Twenty-four points are available, for instance, for manufacturer use of renewable energy for the facility, including purchase of Green-e certified power, and 30 points are available for biobased or recycled content. This means that a 100% recycled-content product made with purchased Green-e power could earn a SMaRT Gold score even if it is deficient in other areas, such as chemical toxicity or greenhouse gas emissions.

Scientific Certifications System (SCS)
Environmentally Preferable Products (EPP)

SCS's multi-attribute EPP certification has been around since 1998. SCS requires products seeking certification to complete a comprehensive environmental performance declaration. Using data gathered through the declaration process, products are certified in one of following ways:

1. A declaration of reduced impact signifies that a product has demonstrated reduced environmental impact compared to the industry baseline for each impact measured.
2. An Environmentally Preferable Product meets additional thresholds for impact reduction relative to the average for its category. Thresholds are determined by a stakeholder consensus process.

NSF-140 and -332

Under the auspices of NSF International, the carpet industry was the first to produce an ANSI-approved, multi-attribute, third-party standard for environmentally preferable building materials. Carpet manufacturers have their products third-party certified to NSF-140 (NSF/ANSI 140-2009 Sustainable Assessment for Carpet), which has been widely adopted by the industry as a leadership standard since its original launch in 2007. NSF-332 covers a range of flooring products, including vinyl composition tile, vinyl and rubber sheet flooring, and linoleum sheet flooring and tile. The standards are point-based and provide flexibility in allowing manufacturers to choose which points to achieve, but they also include additional prerequisites at the higher certification levels.

FloorScore

FloorScore is a third-party indoor air quality certification and label program for low-emitting and hard-surface flooring products, including adhesives. The Resilient Floor Coverings Institute (RFCI) adopted the emissions certification program from the Scientific Certification Systems (SCS), which serves as the third-party certifier. The SCS periodically visits manufacturing plants to verify product content and manufacturing procedures.

Sustainable Attributes Verification and Evaluation (SAVE) Program

SAVE is a third-party, single- and multi-attribute program where manufacturers can have their products evaluated by the ICC Evaluation Service, Inc. (ICC-ES). SAVE provides an independent and comprehensive evaluation of the sustainable attributes of building products. A SAVE evaluation includes product testing at recognized laboratories (when applicable), inspection of the manufacturer's production process, and evaluation of data justifying identifiable attributes.

The ICC-ES has developed evaluation guidelines for use in verifying manufacturers' claims about specific sustainable attributes of their products. These guidelines address the entire production process beginning with raw material acquisition and progressing through final manufacturing and packaging (cradle to gate). Evaluation guidelines include determinations for recycled content, biobased materials, solar reflective index (SRI), volatile organic compounds (VOCs), formaldehyde emissions, and certified woods.

GREEN BUILDING RATING SYSTEMS

LEED-NC

LEED-NC is a rating system composed primarily of elective provisions that are assigned points. LEED-NC contains few mandatory provisions. Its mandatory provisions are its "prerequisites." The intent of LEED-NC's Indoor Environmental Quality section is to improve the health and well-being of building occupants by providing low-emitting materials, reduced material exposure to moisture, increased ventilation and air filtration, pollutant source control, views to the outdoors, daylighting, and adaptability of the thermal and lighting

Table 9-14
LEED for New Construction: Indoor Environmental Quality

Credit	Checklist Prerequisites and Credits
EQp1	Minimum IAQ Performance
EQp2	Environmental Tobacco Smoke (ETS) Control
EQc1	Outdoor Air Delivery Monitoring
EQc2	Increased Ventilation
EQc3.1	Construction IAQ Management Plan, During Construction
EQc3.2	Construction IAQ Management Plan, Before Occupancy
EQc4.1	Low-Emitting Materials, Adhesives, and Sealants
EQc4.2	Low-Emitting Materials, Paints, and Coatings
EQc4.3	Low-Emitting Materials, Flooring Systems
EQc4.4	Low-Emitting Materials, Composite Wood, and Agrifiber Products
EQc5	Indoor Chemical, and Pollutant Source Control
EQc6.1	Controllability of Systems, Lighting
EQc6.2	Controllability of Systems, Thermal Comfort
EQc7.1	Thermal Comfort, Design
EQc7.2	Thermal Comfort, Verification
EQc8.1	Daylighting and Views, Daylight 75% of Occupied Spaces
EQc8.2	Daylighting and Views, Views for 90% of Occupied Spaces

environment for occupant control. There are 2 prerequisite and 17 credits (see Table 9-14).

LEED for Homes

LEED for Homes is a rating system composed primarily of elective provisions that are assigned points. LEED-NC contains few mandatory provisions. Its mandatory provisions are its "prerequisites." Similar to LEED-NC, the intent of LEED for Home's Indoor Environmental Quality section is to improve indoor air quality by providing outdoor ventilation and air filtration, exhausting indoor pollutants, and minimizing indoor moisture. There are 7 prerequisites and 18 optional credits (see Table 9-15).

Green Globes

Green Globes is a rating system containing elective provisions that are assigned point values. Green Globes does not contain mandatory requirements. The Green Globes Indoor Environment category assessment consists of online survey questions organized under the following six stages of the design construction process: (1) Pre-design, (2) Schematic Design, (3) Design Development, (4) Construction

Table 9-15
LEED for Homes: Indoor Environmental Quality

Credit	Checklist Prerequisites and Credits
EQc1	ENERGY STAR with Indoor Air Package
EQp2.1	Basic Combustion Venting Measures
EQc2.2	Enhanced Combustion Venting Measures
EQc3	Moisture Load Control
EQp4.1	Basic Outdoor Air Ventilation
EQc4.2	Enhanced Outdoor Air Ventilation
EQc4.3	Third-Party Performance Testing
EQp5.1	Basic Local Exhaust
EQc5.2	Enhanced Local Exhaust
EQc5.3	Third-Party Performance Testing
EQp6.1	Heating and Cooling Room-by-Room Load Calculations
EQc6.2	Return Air Flow / Room-by-Room Controls
EQc6.3	Third-Party Performance Test / Multiple Zones
EQp7.1	Good Air Filters
EQc7.2	Better Air Filters
EQc7.3	Best Air Filters
EQc8.1	Indoor Contaminant Control during Construction
EQc8.2	Indoor Contaminant Control
EQc8.3	Pre-Occupancy Air Flush
EQp9.1	Radon-Resistant Construction in High Risk Areas
EQc9.2	Radon-Resistant Construction in Moderate Risk Areas
EQp10.1	No HVAC in Garage
EQc10.2	Minimize Pollutants from Garage
EQc10.3	Exhaust Fan in Garage
EQc10.4	Detached Garage or No Garage

Documents, (5) Contracting/Construction, and (6) Commissioning. Each stage assesses indoor environment priorities with the objective of reducing environmental and health impacts (see Table 9-16).

CODES AND STANDARDS

GBI-01 Green Building Assessment Protocol for Commercial Buildings

GBI-01 is a rating system. It is also an ANSI-approved standard. Chapter 12 of GBI-01 addresses the indoor environment. GBI-01 requires that at least 32% of the 160 available points in the indoor environment chapter be satisfied. See Table 9-17 for a summary of the Chapter 12 GBI-01 provisions.

Table 9-16
Green Globes Indoor Environment Assessment by Project Stage

Pre-Design Stage
- Objective to provide a healthy environment for occupants
- Objective to provide an environment that enhances occupant well-being

Schematic Design Stage
- Strategies for effective ventilation
- Strategies for the source control of indoor pollutants

Design Development Stage
- Strategies for effective ventilation
- Strategies for the source control of indoor pollutants

Construction Documents Stage
- Ventilation

Contracting and Construction
- Provision of a healthy environment
- Minimizing disturbance to neighbors during construction

Commissioning
- Commissioning of ventilation
- Commissioning of systems that could produce microbial contamination
- Commissioning of lighting and controls
- Commissioning of thermal control systems
- Commissioning of acoustic and vibration control features

Table 9-17
GBI-01 2010 Green Building Assessment Protocol for Commercial Buildings

Chapter 12 –Indoor Environment

Ventilation Systems

10 points awarded where the quantity of ventilation air is in accordance with ASHRAE 62.1-07 or the 2009 International Mechanical Code.

10 points awarded where air exchanges for mechanically or naturally ventilated buildings are in accordance with 12.1.12.1.

8 points awarded where ventilation intakes and exhausts are equipped with features in accordance with 12.1.3.1.

6 points awarded where CO_2 sensing and ventilation control equipment is provided in accordance with 12.1.4.1.

5 points awarded where air handling equipment measures are implemented in accordance with 12.1.5.

Source Control of indoor Pollutants

2 to 12 points awarded where volatile organic compounds (VOCs) are controlled in accordance with 12.2.1.1.

5 points awarded where HVAC systems address indoor dew point, materials are resistant to mold growth and floor drains are provided at equipment in accordance with 12.2.2.1.

4 points awarded where access is provided for HVAC system maintenance in accordance with 12.2.3.1.

3 points awarded where carbon monoxide monitors are provided in accordance with 12.2.4.1.

2 to 4 points awarded where wet cooling tower drift eliminators and inlet air louvers are in accordance with 12.2.5.1.

2 points awarded where domestic hot water systems maintain hot water temperatures in accordance with 12.2.6.1.

Table 9-17 *(Continued)*

2 points awarded where humidification and dehumidification systems are provided with drain pans in accordance with 12.2.6.1.

1 point awarded where steam humidification systems or ultrasonic humidification systems are provided.

1 point awarded where separate ventilation systems or isolation is provided for specific activities in accordance with 12.2.8.1.

1 point awarded where negative pressure is provided for specific activities in accordance with 12.2.8.2.

Source Control

2 points awarded where integrated pest and control strategies are implemented in accordance with 12.3.1.1.

2 points awarded where sealed storage area is provided for food/kitchen waste in accordance with 12.3.1.2.

Radon Entry and Control

2 points awarded where radon potential is mitigated in accordance with 12.3.2.1.

Lighting Design and Integration of Lighting Systems

3 to 11 points awarded where primary occupied spaces are provided with daylight in accordance with 12.4.1.1.

3 to 9 points awarded where interior spaces are provided with views in accordance with 12.4.1.2.

2 to 6 points awarded where shading devices or photo-responsive controls to maintain consistent lighting levels are provided in accordance with 12.4.1.3.

Lighting Design

7 points awarded where lighting levels for primary occupied spaces are in accordance with 12.4.2.1.

2 to 6 points awarded where reflective glare on visual display terminals is reduced in accordance with 12.4.2.2.

Thermal Comfort

5 to 10 points awarded where thermal control zones are provided for occupancies in accordance with 12.5.1.1.

10 points awarded where the building is in accordance with ASHRAE 55-04.

Acoustic Comfort

4 points awarded where interior sound control measures are implemented in accordance with 12.6.1.1.

2 points awarded where sound transmission ratings of building walls, floors and ceilings are in accordance with 12.6.1.2.

2 points awarded where all floor-ceiling assemblies have an impact rating in accordance with 12.6.1.3.

2 points awarded where reverberation time in quiet areas and areas where speech intelligibility is important is in accordance with 12.6.1.4.

Mechanical, Plumbing and Electrical Systems

4 points awarded where mechanical systems background noise is reduced in accordance with 12.6.2.1.

2 points awarded where airborne noise from HVAC systems are reduced in accordance with 12.6.2.2.

2 points awarded where HVAC system structure-borne noise is reduced in accordance with 12.6.2.3.

2 points awarded where plumbing system noise is reduced in accordance with 12.6.2.4.

2 points awarded where electrical system noise is reduced in accordance with 12.6.2.5.

ICC 700 National Green Building Standard

ICC 700 is a rating system. It is also an ANSI-approved standard. Chapter 9 of ICC 700 addresses indoor environmental quality. As a rating system, ICC 700 contains many electives and few mandatory requirements. Most mandatory provisions in ICC 700 are requirements contained in the International Residential Code and no points are typically awarded for such mandatory provisions, though there are exceptions. The standard does, however, require increased points for IEQ at each of its performance levels. A minimum of 36 points at the Bronze, 65 points at the Silver, 100 points at the Gold and 140 points at the Emerald performance levels must be acquired from Chapter 9 in order to comply with the standard. See Table 9.18 for a summary of these provisions.

Table 9-18
ICC 700-2008 National Green Building Standard

Chapter 9 – Indoor Environmental Quality

Pollutant Source Control

Space and Water Heating Options

- 5 points awarded where natural draft water or space heating equipment is not located in conditioned spaces.
- 5 points awarded where return ducts are not located in a garage or are located in mechanical rooms that comply with 901.1.2.
- 3 to 13 points awarded where furnaces, boilers and water heaters are installed within conditioned space.
- 2 to 5 points awarded where heat pump air handlers are provided and located in accordance with 901.1.4.

Fireplaces and Fuel-Burning Appliances

- 7 points awarded where natural gas or propane fireplaces comply with 901.2.1(1)
- 4 points awarded where wood burning fireplaces are in accordance with 901.2.1(2)(a)
- 6 points awarded where factory-built wood-burning fireplaces are in accordance with 901.2.1(2)(b)
- 6 points awarded where wood stove and fireplace inserts are in accordance with 901.2.1(2)(c)
- 6 points awarded where pellet stoves and furnaces are in accordance with 901.2.1(2)(d)
- 6 points awarded where masonry heaters are in accordance with 901.2.1(2)(e)
- 7 points awarded where fireplaces, pellet stoves, masonry heaters or woodstoves are not installed.

Garages

- 2 points awarded where doors in walls common to the garage and dwelling unit are in accordance with 901.3(a). (Mandatory)
- 2 points awarded where a continuous air barrier is installed at walls and ceilings common to the garage and conditioned spaces in accordance with 901.3(b). (Mandatory)
- 4 points awarded where a garage exhaust fan is provided in accordance with 901.3(a)
- 10 points awarded where a garage is not provided or is detached in accordance with 901.3(2).

Wood Materials

- 2 to 10 points awarded where at least 85% of particleboard, MDF, plywood or composite wood products are in accordance with 901.4.

Carpets

- Carpeting must not be installed adjacent to bathing fixtures and water closets.
- 2 to 10 points awarded where at least 85% of carpet, cushion and carpet adhesives are in accordance with 901.5

Hard-Surface Flooring

- 6 points awarded where emissions for at least 85% of hard-surface flooring is in accordance with 901.6.

Wall Coverings

- 4 points awarded where emissions for at least 85% of wall coverings comply with 901.6.

Architectural Coatings

- 5 points awarded where at least 85% of site-applied coatings are in accordance with 901.8.1.
- 8 points awarded where emissions for site-applied interior products are in accordance with 901.8.2.

Adhesives and Sealants

- 5 to 10 points awarded where at least 85% adhesives and sealants are in accordance with 901.9.

Table 9-18 (*Continued*)

Cabinets

- 2 to 6 points awarded where kitchen and bath vanity cabinets comply with 901.10.

Insulation

- 1 to 5 points awarded where duct insulation emissions are in accordance with 901.11.

Carbon Monoxide Alarms

- 3 points awarded where carbon monoxide alarms comply with 901.12.

Building entrance pollutant control

- 1 to 2 points awarded where exterior or interior mats or grilles are installed in accordance with 901.13.

Non-smoking areas

- 1 point awarded where all common areas are designated as non-smoking.

Pollutant Control

Spot Ventilation

- 8 points awarded where kitchen exhaust and range hoods are vented to the outdoors in accordance with 902.1.1(3).
- 2 to 9 points awarded where bath or laundry exhaust fans are provided with automatic timers or humidistats in accordance with 902.1.2.
- 8 points awarded where kitchen range, bath and laundry room exhausts ventilation rates are verified in accordance with 902.1.3.
- 2 to 6 points awarded for ENERGY STAR compliant exhaust fans in accordance with 902.1.4.

Building Ventilation Systems

- 8 to 17 points awarded where a whole building ventilation system that complies with 902.1 is provided.
- 8 points awarded where building ventilation airflow is tested in accordance with 902.2.2.
- 3 points awarded where MERV filters are provided for HVAC equipment in accordance with 902.2.3.

Radon Control

- 10 to 15 points awarded for radon control systems installed in accordance with 902.3.

HVAC System Protection

- 3 points awarded where HVAC supply registers, grilles and terminations are protected or cleaned in accordance with 902.4.

Central Vacuum System

- 5 points where a central vacuum system is provided in accordance with 902.5.

Living Space Contaminants

- 2 to 8 points awarded where the living space is sealed to eliminate contaminants in accordance with 902.6.

Moisture Management: Vapor, Rainwater, Plumbing, HVAC

Capillary Breaks

- 0 points awarded for mandatory requirements
- 3 points awarded where a capillary break is provided between footings and foundation walls.
- 6 points awarded where a vapor retarder is provided on crawlspace floors in accordance with 903.3.1.
- 8 to 10 points awarded where conditioned crawlspaces are sealed to prevent outside air infiltration in accordance with 903.4.

Moisture Control Measures

- 2 to 10 points awarded where moisture control measures in accordance with 903.4.1 are implemented.

Plumbing

- 2 to 9 points awarded where plumbing lines are installed in accordance with 903.5 through 903.5.3.

(Continues)

Table 9-18 *(Continued)*

Duct Insulation

• 2 points awarded where duct insulation is provided in accordance with 903.6(2).

Relative Humidity

• 8 points awarded where indoor relative humidity is controlled in accordance with 903.7.

Innovative Practices

Humidity Monitoring System

• 2 points where a humidity monitoring system in accordance with 904.1 is provided.

Kitchen Exhaust

• 2 points awarded where kitchen exhaust fan ventilation rates and makeup air is in accordance with 904.2.

ASHRAE 189.1 Standard for the Design of High-Performance Green Buildings

Section 8 of ASHRAE 189.1 provides requirements for indoor air quality (IAQ), thermal comfort, acoustical control, daylighting, and low-emitting materials. The standard contains three critical IAQ components: dilution, air cleaning, and source control. ASHRAE Standard 62.1 Ventilation for Acceptable Indoor Air Quality serves as a foundation with ASHRAE 189.1 setting a higher threshold. ASHRAE 189.1 includes a set of mandatory requirements that must be met for all projects, and an option for either a prescriptive or performance set of requirements to demonstrate compliance.

Contaminant Dilution through Ventilation

ASHRAE 189.1 requires that minimum ventilation rates be determined in accordance with the Ventilation Rate Procedure of ASHRAE 62.1. Outdoor air monitoring equipment is required with a measuring device capable of sending an alarm or control signal when the outdoor airflow falls below the minimum airflow rate requirement.

Air Cleaning

Particulate filters must have a minimum Minimum Efficiency Reporting Value (MERV). When the building is located where the air quality is designated to be in non-attainment with the EPA's National Ambient Air Quality Standards for Fine Particulate Matter (PM2.5), filtration with a minimum MERV of 13 is required for all outdoor air intakes.

Source Control

Environmental Tobacco Smoke: Smoking inside a building space is prohibited and is not allowed within 25 ft (7.5 m) of entrances, outdoor air intakes, and operable windows.

Building Entrances: A significant source of building contaminants is foot traffic. The standard requires a three-part building mat system at each entrance. The mat system must consist of separate scraper, absorption, and finishing surfaces (see Figure 9-9). An exception is allowed for individual dwelling units in residential projects.

Figure 9-9
Three-part building entry mat system.

Finishing mats

Absorption mats

Scrapping mats

© CENGAGE LEARNING 2012

Radon Protection: A soil gas retarding system is required to be installed for building projects located on either brownfield sites or regions that have a significant probability of radon concentrations higher than 4 picocuries per liter. In the United States, these areas are identified as Zone 1 counties on the EPA map of radon zones.

Materials Emissions: Standard 189.1 limits contaminant emissions from construction materials and interior furnishings. The two compliance paths—prescriptive and performance—allow the designer flexibility in materials selection.

Prescriptive Compliance Option

For the following categories of materials, the standard requires product testing every three years, unless certified under a third-party certification program.

Adhesives and Sealants. This section applies to adhesives and sealants used in the building interior and applied on-site. There are two options for demonstrating compliance. The first option requires emissions testing and the second option is based on calculating emissions from the contents of a product. The first option requires products to comply with the emissions requirements of California's CA/DHS/EHLB/R-174 for either office or classroom spaces. The second option requires products to comply with the VOC content requirements of the South Coast Air Quality Management District Rule 1168. Aerosol adhesives must comply with the VOC content listed in the Green Seal Standard GS-36.

Paints and Coatings. Similar to adhesives and sealants, these requirements only apply to paints and coatings applied on the building interior. Paints and coatings must meet the emissions requirements for offices or classroom spaces of California Section 01350 or the VOC content requirements of Green Seal Standard GS-11 for architectural paints and coatings. Clear finishes (such as floor coatings, sealers, and shellacs) must comply with SCAQMD Rule 1113.

Floor Coverings. The emissions requirements of California Section 01350 are used to determine compliance for carpeting and for hard surface flooring in offices and classrooms.

Composite Wood: Wood Structural Panel and Agrifiber Products. This section focuses on limiting the occupant's exposure to formaldehyde. Wood structural panels and agrifiber product used on the interior of the building must contain no added formaldehyde. Given that not all manufacturers have been able to meet this requirement, two exceptions are allowed, both permitting low levels of formal dehyde adhesives:

- Products meeting the California Air Resource Board's regulation, Airborne Toxic Control Measure to Reduce Formaldehyde Emissions from Composite Wood Products (verified through third-party certification); or
- Products meeting the California Section 1350 limits for office or classroom spaces.

Office Furniture and Seating. This section limits the occupant's exposure to numerous VOCs emitted from office furniture systems and chairs. These products must be tested according to ANSI/BIFMA Standard M7.14 and demonstrate compliance with the emissions factors or concentration limits that are listed in Appendix E of ASHRAE 189.1.

Ceiling and Wall Systems. This section applies to materials such as ceiling or wall insulation, ceiling or wall panels, gypsum wall board, and other wall covering. These systems must meet the emission limits for office or classroom spaces specified in California Section 01350.

Performance Compliance Option

The performance path compliance option for IAQ is an alternative to the prescriptive option. Under this option, laboratory-derived emission values from material categories using the criteria set in Normative Appendix F of the standard are modeled using the specific characteristics of the building project to calculate space concentrations for chemicals listed in California Section 01350. The sum of each individual VOC concentration from these materials shall be no more than 100% of the limits listed in California Section 01350.

Thermal Comfort

ASHRAE 189.1 has mandatory requirements to comply with the design and documentation sections of ANSI/ASHRAE Standard 55, Thermal Environmental Conditions for Human Occupancy. There are exceptions for certain space types that require conditions outside the normal thermal comfort range, such as food storage areas and natatoriums.

Acoustics

The standard contains mandatory requirements for acoustical control and treatment. A minimum composite Outdoor-Indoor Transmission Class (OITC) rating of wall and roof-ceiling assemblies is required when a building is exposed to higher noise levels. Noise-sensitive spaces, such as offices and classrooms, require a minimum composite Sound Transmission Class (STC) rating for interior wall and floor-ceiling assemblies.

Exterior sound control requirements set minimum composite OITC (or STC) ratings for walls, roof-ceiling assemblies, and fenestration in buildings located near (within 1,000 ft [300 m]) expressways or busy commercial airports (within 5 mi [8 km]).

Interior sound control STC rating levels are specified for wall and floor-ceiling assemblies, depending on the application (see Table 9-19).

Daylighting

The mandatory requirements for daylighting includes toplighting in large enclosed spaces (20,000 ft^2 [2,000] m^2) or greater) directly under a roof with finished ceiling heights greater than 15 ft (4 m) and with a general lighting having a lighting power density or lighting power allowance greater than 0.5 W/ft^2 (5 W/m^2). The standard requires that a minimum of 50% of the floor area directly under a roof shall be in the daylight zone (see Figure 9-10), with a minimum amount of the daylight zone area being within the toplighting area based on the lighting power density or allowance in the zone. The minimum ratio of toplighting area to daylight zone ranges from 3.6% for zones with lighting power of 1.4 W/ft^2 (14 W/m^2) or more to 3.0% for lighting power between 0.5 and 1.0 W/ft^2 (5 to 10 W/m^2).

The toplighting requirements do not apply to buildings located in the cold Climate Zones of 7 and 8, or where it would interfere with the intended use, such as in auditoria, theaters, museums, places of worship, and refrigerated warehouses.

Table 9-19
Sound Transmission Class Rating Requirements for Interior Assemblies

Application Separating	Minimum STC Rating
Dwelling units, adjacent tenant spaces, adjacent classrooms	50
Hotel, motel rooms: patient rooms in nursing homes or hospitals	45
Classrooms adjacent to restrooms or showers	53
Classrooms adjacent to music or mechanical rooms, cafeterias, gymnasiums, and indoor pools	60

© CENGAGE LEARNING 2012

Figure 9-10

Side and plan view of daylight zone under a skylight.

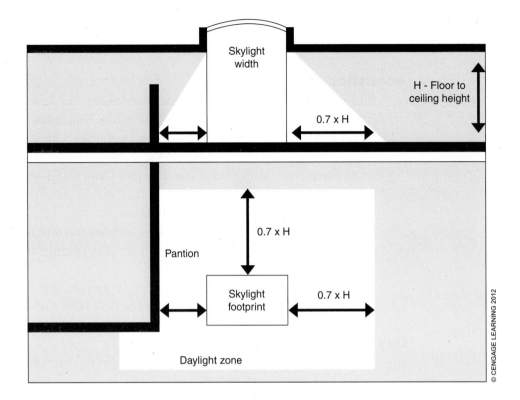

© CENGAGE LEARNING 2012

Under the perspective compliance path, sidelight daylighting is required for office spaces and classrooms. Compliance is determined by the minimum effective aperture for vertical fenestration on north, south, and east wall facades, along with visible light reflectance levels for opaque interior surfaces in the daylight zone.

For west-, south-, and east-facing facades on office spaces, shading is to be provided with devices such as louvers, light shelves, or roof overhangs such that a minimum shading projection factor (PF) of 0.5 is achieved.

Under the performance compliance option, daylighting requirements can be met by daylighting simulation that demonstrates an illuminance of at least 30 footcandles (300 lux) on a plane 3 ft (1 m) above the floor within 75% of the daylight zone area. The simulation must also demonstrate that direct sunlight would not strike anywhere on a work surface for more than 20% of the occupied hours in office spaces. A default value of 2.5 ft (0.75 m) above the floor can be used for the work surface height.

International Green Construction Code (IgCC)

The IgCC is a code composed primarily of minimum baseline requirements. As mentioned in previous chapters, the International Green Construction Code (IgCC) provides a comprehensive set of requirements intended to reduce the negative impact of buildings on the natural environment. Chapter 8 of the IgCC contains requirements to reduce the quantity of building indoor air contaminants and pollutants, including those that are odorous, irritating, or harmful, and to provide an interior environment that is conducive to the health and well-being of building occupants, neighbors, and construction personnel (see Table 9-20).

Table 9-20
IgCC Mandatory IEQ Provisions (Public Version 2.0)

Chapter 8 – Indoor Environmental Quality and Comfort

Section 802: Building Construction Features, Operations and Maintenance Facilitation

802.2 thru 802.5 Requirements to facilitate the operation and maintenance of the completed *building* systems.

- Air handling system access
- Durability and cleanability of air handling surfaces
- Air handling system filters
- Insulation materials above suspended ceilings and in air plenums

Section 803: HVAC Systems

803.1 During the construction phase, duct and other related air distribution component openings shall be covered to reduce collection of dust and debris. Temporary ventilation shall be provided.

803.2 Temperature and humidity in occupied spaces shall comply with ASHRAE 55.

803.3 Environmental tobacco smoke control, containment and ventilation shall be in accordance with this section.

803.4 Isolation of pollutant sources related to print, copy, janitorial rooms and garages shall comply with this section.

803.5 Filters for ducted space conditioning systems shall be rated at MERV 6 or higher and system equipment shall be designed to be compatible.

Section 804: Specific Indoor Air Quality and Pollution Control Measures

804.1 Where located within *buildings, fireplaces*, solid fuel-burning appliances and decorative appliances for installation in *fireplaces* shall comply with this section.

804.2 Buildings in High Radon Potential (Zone 1) locations shall comply with mitigation provisions of this section.

804.3 After construction is completed and all interior finishes are installed, the *building* shall be flushed-out by supplying continuous or intermittent *ventilation* with all air handling units at their maximum outdoor air rate as specified in this section.

Section 805: Asbestos Use Prevention

805.1 The use of and installation of asbestos in *building* construction shall be prohibited by *building* design and construction control measures.

Section 806: Material Emissions and Pollutant Control

806.1 Interior particleboard, hardwood plywood, and medium density fiberboard used as sub-flooring and decorative wall coverings, and in permanently installed millwork, shall be rated for exterior exposure, made using adhesives which do not contain urea-formaldehyde (UF) resins, sealed on all sides and edges, or comply with the requirements for formaldehyde limits as specified in Table 806.1.

806.2 A minimum of 85% by weight or volume, of site applied adhesives and sealants shall comply with the *VOC* content limits in Table 806.2(1) or alternative *VOC* emissions limits in Table 806.2(2). The provisions shall not apply to adhesives and sealants subject to state or federal consumer product *VOC* regulations.

806.3 A minimum of 85% by weight or volume, of site-applied interior architectural paints and coatings shall comply with *VOC* content limits in Table 806.3(1) or the alternate emissions limits in Table 806.3(2).

806.4 A minimum of 85% of the total area of flooring installed within the interior of the *building* shall comply with the VOC emission limits of Table 806.4 (2). Where post manufacture coatings or surface applications have not been applied, the flooring listed in Table 806.4(1) shall be deemed to comply with the requirements of Table 806.4(2).

806.5 A minimum of 85% of acoustical ceiling tiles and wall systems, by square feet, shall comply with the VOC emission limits of Table 806.5(2). Where post manufacture coatings or surface applications have not been applied, the ceiling or wall systems listed in Table 806.5(1) shall be deemed to comply with the requirements of Table 806.5(2).

(Continues)

Table 9-20 (*Continued*)

806.6 A minimum of 85% of insulation shall comply with the VOC emission limits of Table 806.6.
Section 807: Sound Transmission
807.2 Where Groups A1, A3, E, I, B used for educational purposes and R occupancies, are located within 500 feet (150 m) of a roadway containing 4 or more traffic lanes, the wall and roof-ceiling assemblies making up the *building* envelope shall have a sound transmission class (STC) or outdoor-indoor transmission class (OITC) of not less than 50 and the windows within the *building* envelope walls shall have a sound transmission class (STC) or outdoor-indoor transmission class (OITC) of not less than 30.
807.3 Wall and floor-ceiling assemblies that separate Group A, F and M occupancies from Group B, I or R occupancies shall have a sound transmission class (STC) of not less than 50.
807.4 *Building* mechanical and emergency generator systems shall be designed to control airborne *noise* in accordance with Tables 807.4.2 and 807.4.3.
Section 808: Daylighting
808.2 Daylighting of *building* spaces shall be required for buildings containing Group A-3, B, E, F, M or S occupancies in accordance with Section 808.3 with specified exceptions.
808.3 Not less than 50% of the total floor area in *regularly occupied spaces* shall be located within a *daylit area* that complies with either: • Prescriptive provisions of Section 808.3.1 • Performance provisions of Section 808.3.2

BEST PRACTICES FOR IEQ

Indoor environmental quality (IEQ) involves fresh air ventilation exchange, low-emitting materials and finishes, daylighting, visual connection to nature, and accessible controls for lighting and temperature. Strategies for IEQ include:

1. Source Control
   a. Limit entry of outdoor contaminants
   b. Limit contaminants from indoor sources
   c. Capture and exhaust contaminants from building equipment
2. Ventilation
   a. Reduce contaminant concentrations through ventilation, filtration, and air cleaning
   b. Control moisture and contaminants in mechanical systems
3. Connection to Outdoors
   a. Capture daylight for visually oriented tasks and activities
   b. Incorporate visual connection to nature
4. Building Commissioning
   a. Manage the design, construction, and commissioning for IAQ
5. Facilitate Effective Operation and On-going Maintenance

More is not necessarily better and this is particularly true for outside ventilation and daylighting. Over ventilation will jeopardize comfort and increase heating and cooling costs. Excessive daylight produces glare and will negatively impact cooling costs. Building design must establish a balance between IEQ, energy efficiency and occupant comfort.

Finally, IEQ goals and strategies should be set early in the schematic phase of building design, which requires a coordinated effort between the architect; mechanical, structural, and electrical engineers; and interior and landscape designers. An IEQ coordinator should be part of the project design team and have the responsibility for coordination between the disciplines to ensure IEQ goals are met through design, construction, and building commissioning.

BUILDING COMMISSIONING, OPERATION, AND MAINTENANCE

THE IMPORTANCE OF PLANNING, OPERATIONS, AND MAINTENANCE FOR BUILDING SUCCESS

Building commissioning, operation, and maintenance are imperative to the success of any building and particularly for green buildings. Building commissioning provides the owner with an unprecedented level of assurance that the building will function as designed with assured reliability and reduced operating costs. Operation and ongoing maintenance following the basis of design will reduce equipment downtime, increase building system life, and contribute to occupant health and worker productivity. The proper training of the owner's operations and maintenance staff is critical to this success.

SURVIVAL OF THE FITTEST—DESIGN AND CONSTRUCTION

Integrated Project Delivery

The design and construction phase is the time to set the stage for the operations and maintenance of building systems. Integrated project delivery (IPD) is an effective contract tool designed to minimize the shortcomings of coordination between design and building project team members. Like other building project delivery methods, such as design-bid-build and design-build, IPD is a legal framework that the owner, the architect, and the builder can use to collaborate on a project. IPD is "a project delivery approach that integrates people, systems, business structures and practices into a process that collaboratively harnesses the talents and insights of all participants to reduce waste and optimize efficiency through all phases of design, fabrication and construction."[1]

The design-bid-build delivery method is the most common contract mechanism. It's primary objective is low cost delivery of the project and involves three primary parties: owner, designer, and builder. Those parties form two separate contracts, one between the owner and the designer, and another between the owner and the contractor. Projects typically proceed in a linear fashion from designing to bidding to building, with the builder getting involved only after the design team has fully developed the design and construction documents. A general contractor is selected by submitting the lowest bid, offering the best value, or by negotiating the best package. Similarly, the general contractors selects subcontractors based on competitive bidding.

---

[1]AIA California Council, *Integrated Project Delivery* (McGraw Hill Construction, 2007).

Although the project is theoretically delivered at the lowest cost to the owner, conflicts among the parties to the contract are common. In part because of the contractor's lack of involvement in the preparation of the construction documents, those documents often lack details necessary for the contractor to meet the design intent. Delays and breakdown in the productive communication leads to higher costs from change orders, repairs, and lawsuits.

The design-build approach is an alternative delivery method to design-bid-build. The owner's risks can be reduced by dealing with a single contract with a firm combining design and construction services. This will reduce conflicts, facilitate better communication, and improve outcomes. In addition, the project schedule is usually shortened due to the overlapping phases of design and construction. Barriers to integrated design are reduced because the single firm combines multiple functions. In addition, the firm's interest is more in line with the owner's for a productive outcome.

Design-build has been limited depending on the scope of a project and the degree to which design-build firms are willing to take on the risk of an entire project. Buildings such as warehouses and schools benefit from the efficiencies of design-build, whereas projects with more complex designs and building systems may benefit less. Also, depending on how a project contract is structured, the design-build firm can be rewarded by reducing construction costs instead of seeking project improvements that better meet the owner's objectives.[2]

Another alternative to design-bid-build is construction management-at-risk. Under this delivery method, the owner contracts separately with the design team and a construction management firm who serves as the owner's representative. Early in the design process, the construction manager is required to guarantee a total construction cost for the project, which is referred to as the guaranteed maximum price (GMP). Conflict and miscommunication are minimized because the design team works closely with the construction manager in preparing the design and construction documents that meet the owner's requirements and resolve potential construction detail concerns.[3]

IPD leverages knowledge and expertise through early collaboration and utilization of new technologies, allowing all team members to better realize their potentials. At the core of an integrated project is a collaborative and integrated team composed of key project participants. The team is guided by transparency, collaboration, information sharing, shared risk and reward, and use of full technological capabilities and support. The outcome is an optimal environment in which to design, build, and operate as efficiently as possible. Table 10-1 compares traditional project delivery and integrated project delivery.[4]

IPD moves design and construction decisions to earlier in a project in comparison to conventional project delivery methods. This helps ensure positive building

---

[2]Altes, Tristan Korthals, "Intergrated Project Delivery: A Platform for Efficienct Construction," *Environmental Building News* (Vol. 17 No. 11, November 2008)

[3]Kibert, Charles J., *Sustainable Construction: Green Building Design and Delivery* (Hoboken, NJ: Wiley, 2008)

[4]AIA and AIA California Council, *Integrated Project Delivery: A Guide* (Washington, DC: AIA,2007)

Table 10-1

Traditional versus Integrated Project Delivery

Project Element	Traditional	Integrated
Fragmented, assembled on "just-as-needed "minimum-necessary" basis, strongly hierarchical, controlled	Teams	An integrated team entity composed of key project stakeholders, assembled early in the process, open, collaborative
Linear, distinct, segregated; knowledge gathered "just—as-needed", information hoarded; silos of knowledge and expertise	Process	Concurrent and multi-level; early contributions of knowledge and expertise; information openly shared; stakeholder trust and respect
Individually managed, transferred to the greatest extent possible	Risk	Collectively managed, appropriately shared
Individually pursued; minimum effort for maximum return; (usually) first-cost based	Compensation/Reward	Team success tied to project success; value-based
Paper-based, two-dimensional; analog	Communications/Technology	Digitally based, virtual; Building Information Modeling (three-, four-, and five-dimensional)
Encourage unilateral effort; allocate and transfer risk; no sharing	Agreement	Encourage, foster, promote, and support multi-lateral open sharing and collaboration; risk sharing

outcomes while reducing construction delays and overall project costs. Decisions made later in a project tend to cost more while having less overall benefit.[5] The "MacLeamy Curve" shows how integrated project delivery (IPD) moves design and construction decisions to earlier in a project.

The following are common features of IPD:[6]

- **Mutual respect and trust:** Parties work in a collaborative environment. The emphasis is on communication to solve problems rather than assign blame.
- **Mutual benefit and reward:** Compensation typically rewards early and active involvement of the architect and the builder, with incentives for working in the best interests of the project.
- **Collaborative innovation and decision making:** The project team makes decisions under consensus, and parties work across boundaries to spur innovation.
- **Early involvement of key participants:** The architect and builder are typically hired at the same time at the start of a project.
- **Early goal definition:** The team formulates specific project outcomes. Effort is "front-loaded" during the design phase and green design is often explicitly integrated.
- **Appropriate design technology:** Project teams often use advanced design technologies such as Building Information Modeling (IBIM) to maximize collaboration, efficiency, interoperability, and transparency.

[5]AIA and AIA California Council, *Integrated Project Delivery: A Guide* (Washington, DC: AIA,2007)

[6]Altes, Tristan Korthals, "Intergrated Project Delivery: A Platform for Efficienct Construction," *Environmental Building News* (Vol. 17 No. 11, November 2008)

Integrated project delivery principles can be applied to a variety of contractual arrangements. At a minimum, though, an integrated project requires a close collaboration between the owner, architect, and the general contractor responsible for construction of the project, from early design through project building commissioning. New contractual arrangements will need to occur to achieve these outcomes, including insurance and legal arrangements.

While integrated design is well suited to large projects for the coordination of design teams, it can be burdensome and inefficient for small to medium size

Project Delivery Methods

Design-Bid-Build (Hard Bid)

- The primary objective of a design-bid-build delivery system is the low-cost delivery of a project. The design team is selected by the owner and works on the owner's behalf to produce construction documents. General contractors bid on the project, with the lowest qualified bidder getting the job. The general contractor in turn selects subcontractors based on competitive bidding and awards the work to the lowest qualified bidder. Conflicts and miscommunication among parties are common and often result in higher costs from change orders and repairs.

Design-Build

- A project delivery method in which one entity (design/builder) executes a single contract with the owner to provide design and construction services. This entity may be a single firm that provides both design and construction services or a partnership between a design firm and construction company. This arrangement is more likely to reduce the typical design/construction conflicts, improve communication among the team members, reduce overall cost to the owner, improve quality, and speed the process.

Construction Management-at-Risk (Negotiated Work)

- The owner contracts separately with the design team and the contractor or construction manager, who works on the owner's behalf. Early in the design process, the construction manager is required to guarantee that the total construction costs will not exceed a maximum price. The construction manager provides preconstruction services, including cost analysis, constructability, value engineering, and project scheduling, based on the design team's drawings. Working together, the parties produce construction documents that meet the owner's requirements, schedule, and budget, and minimize conflicts and miscommunication. The construction manager selects subcontractors based on their capabilities and not merely the lowest bid.

projects. For such projects, a preferred method is that the project manual contain requirements that are well conceived and written to ensure that, when complied with, the result is a design that has inherently given attention to the same details and goals that are intended to be addressed and achieved through integrated design strategies.

Advancements in computer aided drafting and design and building information modeling have made it possible for a single designer to produce effective and complete construction documents for small to medium projects. In many cases, a competent designer can feasibly achieve the same results as a large and complex design team, and do so much more cost effectively. The bottom line is that the results should be the most important factor, not necessarily the method used to achieve them.

Building Commissioning

With ever-increasing advancements in technologies, buildings contain sophisticated systems for controlling comfort, lighting, equipment, providing security from intruders, and enhancing occupant safety. Building envelope and structural systems are also more complex, with engineered membranes, coatings, and spectral-selective glazing systems. In Figure 10-1, complex building systems in this mixed-use, 700-unit, 11-building, mixed-use condominium project include multistory cantilevered vegetated roof terraces, high-performance glazing, two-level interconnected underground parking structure, and a stormwater management filtration system with rainwater collection tanks.

Figure 10-1

Complex building systems require an integrated design, project delivery, and commissioning process.

Designing and installing these systems requires specialists who may have only a limited understanding about how each system affects other aspects of the building. Design team members are often constrained by tight budgets and liability concerns about system components that are not within their area of expertise or responsibility. It is not unusual for large, complicated buildings to be built without any one party having overall responsibility for ensuring that the systems function properly together.

The purpose of building commissioning is to provide documented confirmation that building systems function in compliance with criteria set forth in the project documents to satisfy the owner's operational needs.[7] Commissioning usually involves an independent party, other than the designer or contractor, who is charged with ensuring that a completed building meets the design intent and requirements of the owner. Commissioning is a comprehensive version of standard mechanical system testing, adjusting, and balancing (TAB). Besides the heating, ventilation, and air-conditioning (HVAC) system, commissioning includes other building systems, from lighting and renewable energy systems to fire safety and in some cases building structure, exterior envelope, and finishes.

The role of the commissioning agent begins earlier in the design process, with the goal of identifying and resolving potential issues before the construction phase. Commissioning agents review design development drawings, construction documents and subsequent change orders and submittals from contractors that affect the systems being commissioned. At the conclusion of construction, a commissioning agent not only verifies and tests building components and systems but also provides a final report that organizes for the owner many important details about operational guidelines and maintenance requirements. Often the commissioning agent will continue to monitor the building for the first year or two of operation, helping to ensure that operations and maintenance (O&M) procedures do not undermine system performance, and helping to resolve problems that arise. The commissioning agent can serve as the owner's representative from the earliest stages of the design process through building occupancy, helping to define design goals and performance criteria, and then monitoring to see that those goals are achieved.

Building commissioning consists of the following primary tasks:[8]

1. Documentation of design intent.
2. Field-testing of systems and equipment.
3. Transfer of operation and maintenance requirements and supporting documents to building management.

Building commissioning requires a commitment on behalf of the owner, project team, and designated commissioning authority. Table 10-2 lists the tasks and shared responsibilities.

---

[7] Building Commissioning Association, http://www.bcxa.org/.

[8] Grondzik, Walter, *Principles of Commissioning* (Hoboken: John Wiley & Sons, 2009).

Table 10-2
Task and Responsibilities for Building Commissioning[9]

Project Phases	Commissioning	Pre-Occupancy	Post-Occupancy
Predesign, Design Phase			
Request for proposal architect and engineer selection	1. Designating commissioning authority (CxA)	Owner or project team	N/A
Owners project requirements (OPR); basis of design (BOD)	2. Document owner's project requirements; develop basis of design	Owner or CxA and project team	N/A
Schematic design	3. Review owner's project requirements and basis of design	Project team or CxA	N/A
Design development	4. Develop and implement commissioning plan	Project team or CxA	N/A
Construction documents	5. Incorporate commissioning requirements into construction documents 6. Conduct commissioning design review prior to mid-construction documents	Project team or CxA	N/A
Construction Phase			
Equipment procurement and installation	7. Review contractor submittals applicable systems being commissioned	CxA	N/A
Functional testing; test and balance; performance testing acceptance	8. Verify installation and performance of commissioned systems	CxA	CxA
Operations and maintenance (O&M) manuals	9. Develop systems manual for commissioned systems	Project team or CxA	N/A
O&M training	10. Verify that requirements for training are completed	Project team or CxA	Project team or CxA
Substantial completion	11. Complete a summary commissioning report	CxA	N/A
Occupancy			
Systems monitoring	12. Review building operation within 12 months after occupancy	N/A	CxA

Although the commissioning of mechanical systems is the primary commissioning activity, the commissioning process should encompass all building systems, including lighting controls, electrical components, on-site renewable energy

---

[9]Adapted from USGBC, *LEED Green Building Design and Construction Reference Guide* (Washington, DC: USGBC, 2009)

Figure 10-2

Renewable energy system.

COURTESY OF ANTHONY FLOYD

generation systems, monitoring systems, tele-communications, fire protection and security systems, electrical water controls, graywater systems, and rainwater harvesting systems (see Figures 10-2 and 10-3).

While commissioning has traditionally been thought of as primarily related to energy, green codes, standards and rating systems have placed as much emphasis on other areas, including water and material conservation, indoor environmental quality and site use and development. Table 903.1 of the IgCC is a good example of the many areas that commissioning can address. It is also important to note that building officials who perform field inspections are, in fact, actually performing commissioning related activities.

BUILDING OPERATIONS AND MAINTENANCE

To ensure that building systems operate optimally for the life of the building, operation and maintenance documents must be developed to help building owners and facility managers operate the building consistent with the design intent and equipment specifications. The documents should include the following:[10]

1. **Building operating plan:** Defines the conditions required by building management and occupants for successful operations of the building, including desired temperature, humidity levels, fresh air ventilation, lighting levels, and time-of-day schedules.
2. **Systems narrative:** A summary description of building systems, including heating, cooling, ventilation, water heating, humidification and/or dehumidification, lighting, and associated control systems.
3. **Sequence of operations:** Defines operational states desired for different conditions, including operations for full or partial loads, staging or cycling of equipment, water temperatures, air pressures, and operating phases for warm-up, occupied, or unoccupied conditions.
4. **Preventive maintenance plan:** Should include the manufacturer's recommendations for the ongoing operation of the base building systems.

---

[10]USGBC, *LEED Green Building Design and Construction Reference Guide* (Washington, DC: USGBC, 2009)

Figure 10-3

Rainwater collection allows on-site management of stormwater to meet irrigation needs without potable water.

5. **Commissioning report:** Identifies problems that are likely to emerge or require particular attention on an ongoing basis.

Table 10-3 lists common training topics for the building owner and operational staff to be able to operate the facilities optimally over the useful life of the building.

CODES AND STANDARDS

GBI-01 Green Building Assessment Protocol for Commercial Buildings

GBI-01 is a rating system. It is also an ANSI-approved standard. Chapter 6 of GBI-01 contains provisions that encourage the implementation of commissioning, operations and maintenance related practices. GBI Chapter 6 also contains other provisions related to project management for green design and delivery coordination. GBI-01 requires that at least 50 percent of the 100 points available in Chapter 6 be satisfied. It is possible, however, to acquire all required points for Chapter 6 for the implementation of practices that are not related to commissioning, operations and maintenance. See Table 10-4 for a summary of the commissioning, operations, and maintenance related provisions of Chapter 6 of GBI-01.

Table 10-3
Common Training Topics

General purpose of systems (design intent)
Use of O&M manuals
Review of control drawings and schematics
Startup, normal operation, shutdown, unoccupied operation, seasonal changeover, manual operation, control setup, programming, and alarms
Interaction with other system
Adjustments and optimizing methods for energy conservation
Health and safety issues
Special maintenance and replacement sources
Occupant interaction issue
System response to different operating conditions

COURTESY OF ANTHONY FLOYD

© CENGAGE LEARNING 2012

Table 10-4
GBI-01 2010 Green Building Assessment Protocol for Commercial Buildings Commissioning, Operations, and Maintenance

Pre-Commissioning
3 points awarded where the owner's project requirements for building systems and the basis of design are documented and an independent commissioning authority reports to the building owner in accordance with 6.3.2.1.

Whole Building Commissioning
5 points awarded where the building envelope is commissioned in the pre-design, design, and construction phases in accordance with 6.3.2.1.
5 points awarded where HVAC&R are commissioned in the pre-design, design, and construction phases in accordance with 6.3.2.2.
4 points awarded where the structural system is commissioned in the pre-design, design, and construction phase in accordance with 6.3.2.3.
4 points awarded where fire protection systems are commissioned in the pre-design, design, and construction phases in accordance with 6.3.2.4.
3 points awarded where plumbing systems are commissioned in the pre-design, design, and construction phases in accordance with 6.3.2.5.
3 points awarded where electrical systems are commissioned in the pre-design, design, and construction phases in accordance with 6.3.2.6.
3 points awarded where lighting systems are commissioned in the pre-design, design, and construction phases in accordance with 6.3.2.7.
2 to 6 points awarded where interior, elevating, conveying, and communication systems are commissioned in the pre-design, design, and construction phases in accordance with 6.3.2.8.
2 points awarded where field testing of partitions for noise isolation are conducted in accordance with 6.3.2.9.
2 points awarded where building system specifications are commissioned in accordance with 6.3.2.10.
2 points awarded where training on commissioning systems is provided in accordance with 6.3.2.11.

Operations and Maintenance Manuals
14 points awarded where an operations and maintenance manual is provided in accordance with 6.4.1.1.

ICC 700 National Green Building Standard

ICC 700 is a rating system. It is also an ANSI- approved standard. Chapter 10 of ICC 700 addresses building operation, maintenance and owner education. As a rating system, ICC 700 contains many electives and few mandatory requirements. The standard does, however, require increased points for operation and maintenance at each of its performance levels. A minimum of 8 points at the Bronze, 10 points at the Silver, 11 points at the Gold and 12 points at the Emerald performance level must be acquired from the ICC 700 Chapter 10. See Table 10-5 for a summary of these provisions. Note that Chapter 7 of ICC 700 contains provisions related to the commissioning of energy systems and equipment. See Chapter 8 of this book for a summary of those ICC 700 energy provisions.

Table 10-5
ICC 700-2008 National Green Building Standard

Chapter 10: Operation, Maintenance, and Building Owner Education
Building Owners' Manual for One- and Two-Family Dwellings
Mandatory Requirements
1) A certificate or completion document from a green building program
2) A list of the buildings green features
3) Product manufacturer's manuals and data sheets for major equipment, appliances, and fixtures
Elective Provisions
1 point awarded for each 2 items included in the owners' manual from the list in 1001.1
Training of Building Owners on Operation and Maintenance for One- and Two-Family Dwellings and Multi-Unit Buildings
6 points awarded where training of building owners with regard to green building goals and strategies implemented in the building is provided in accordance with 1002.1
Construction, Operation and maintenance manuals and Training for Multi-Unit Buildings
Building Construction Manual (5 points max)
Mandatory Requirements:
1) A certificate or completion document from a green building program
2) A list of the buildings green features
3) Product manufacturer's manuals and data sheets for major equipment, appliances and fixtures
Elective Provisions:
1 point, plus an additional awarded for each 2 items complied with, including those items indicated to be mandatory, from the list of items in 1003.1
Operations Manual (5 points max)
Mandatory Requirements:
1) Details listing the importance of operating and living in a green building
2) A list of conservation practices
3) Details of the importance of maintaining proper building relative humidity
Elective Provisions:
1 point, plus an additional point awarded for each 2 items complied with, including those items indicated to be mandatory, from the list of items in 1003.2
Maintenance Manual (5 points max)
Mandatory Requirements:
Documentation of the importance of maintaining a green building.
Elective Provisions:
1 point, plus an additional point awarded for each 2 items complied with, including those items indicated to be mandatory, from the list of items in 1003.2

ASHRAE 189.1

Section 10 of ASHRAE 189.1 specifies requirements during the course of construction of a building and requires plans for building operation. Construction activities addressed include acceptance testing, commissioning, erosion and sediment control, and construction indoor air quality management. The development of plans for operation must include water and energy consumption tracking, IAQ monitoring, maintenance and service life, and transportation. All of the requirements in Section 10 are mandatory, so there are no prescriptive or performance paths.

Acceptance Testing and Commissioning

To ensure optimal building performance is achieved, building systems must be verified to operate as designed. Standard 189.1 includes requirements for acceptance testing of all building projects. An acceptance representative must be designated prior to obtaining a building permit to lead, review, and oversee completion of acceptance testing activities, and the representative must review construction documents to verify measurement sensor locations, devices, and control sequences. Before the building is occupied, acceptance tests must be performed that include sign-offs from all responsible parties, and system manuals must be provided to facility operation and maintenance staff.

A commissioning authority (CxA) must be designated and lead a process documenting the owner's project requirements (OPR), as well as review construction documents throughout the design process. The design team must document a basis of design that reflects the needs developed in the OPR. During building construction, systems must be commissioned to verify performance and conformance to the OPR, and the owner's facility operations staff must be trained as part of the commissioning process.

Systems requiring commissioning include HVAC, IAQ, water heating, building envelope, lighting and shading controls, plumbing, domestic and process water, irrigation, water measurement devices, renewable energy systems, and energy measurement devices. Procedures, documents, tools, and training must be provided to the facility operations and maintenance staff to maintain these systems during the service life of the building.

Erosion and Sediment Control

ASHRAE 189.1 requires the development of an erosion and sediment control (ESC) plan for all construction. The ESC plan must conform to the requirements of the most current version of the EPA National Pollutant Discharge Elimination System (NPDES) or local erosion and sedimentation control standards and codes—whichever is the most restrictive.

IAQ Construction Management

To reduce indoor air quality problems, ASHRAE 189.1 provides two building options.

The first option is to deliver a total volume of outdoor air, measured in total air changes (TAC) to the building based on ANSI/ASHRAE Standard 62.1. Alternatively, as a second option, contaminant testing of HVAC airstreams must demonstrate that concentration levels of specific contaminants do not exceed a set threshold. The threshold levels are based on national standards such as ANSI/ASHRAE 62.1 and California's Department of Health Services Specification 01350.

Permanent HVAC systems are prohibited from being operated during construction except for system startup, testing, balancing, and commissioning, and all filters and controls must be in place and operational during flush out.

Vehicle staging areas must be established for loading or unloading materials. These staging areas must be located 100 feet (30.48 m) from any outdoor air intakes, operable openings, and hospitals, schools, residences, hotels, daycare facilities, elderly housing, and convalescent facilities.

Plans for Operation

The standard requires the project team to develop plans for operation based on the knowledge of system design coupled with the needs of the building owner. The standard does not dictate specific operational requirements but prescribes what the plan must include.

Water and Energy Use

ASHRAE 189.1 requires that the plan for operation include a process for monitoring and assessing energy and water use. The initial assessment must be completed 12 months, but no later than 18 months, after building occupancy; further assessments must be completed at least every three years. The plan must require energy- and water-use reports that include hourly load profile for each day, monthly average daily load profile, monthly and annual energy use, and monthly and annual peak demand.

Water-use and energy-consumption data must be entered into the ENERGY STAR Portfolio Manager to track building performance and provided to and retained by the owner.

Indoor Environmental Quality Plan

The plan for operation must provide an indoor environmental quality measurement and verification plan that includes requirements for monitoring outdoor air flow and indoor air quality measures.

ASHRAE 189.1 also requires a plan for general cleaning during building operation according to Green Seal Standard GS-42. This includes environmentally preferable cleaning products and practices. The standard covers cleaning of entryways, floors, restrooms, and dining rooms; vacuuming; disinfecting; and solid waste, trash collection, and recycling (see Figure 10-4).

Maintenance and Service Life Plan

The plan for operation must include maintenance plans addressing all mechanical, electrical, plumbing, and fire protection systems, and a service life plan addressing structural, building envelope, and hardscape materials. The intent of these plans is to address the long-term impacts of design decisions and to inform facility managers in order to optimize building operations and maintenance.

The service life plan requires identification of materials that need to be inspected, repaired, or replaced during the design life, which is generally at least 50 years. The plan must include the estimated service life, maintenance frequency, and access for maintenance. It is the responsibility of the owner to retain the document during the life of the building.

Figure 10-4

Trash and recycling collection in a multi-level building.

TRASH DISPOSAL GUIDELINES

• Trash must be placed in secure bags.

• Boxes and other items that do not fit in the trash chute must be disposed of at the LL1 trash room.

• Please follow city recycling guidelines posted on city website

Additional information available at the Property Management office

RECYCLE

TRASH

COURTESY OF ANTHONY FLOYD

Transportation Management Plan

Standard 189.1 requires a transportation management plan that includes preferred parking for carpools and vanpools as well as a plan for bicycle transportation for all projects. Owners of a building project who occupy or partially occupy a building must provide employee benefits that either incentivize the use of mass transit, ridesharing, or carpooling or initiate telecommuting or a flexible work schedule programs.

International Green Construction Code (IgCC)

The IgCC is a code composed primarily of minimum baseline requirements. As mentioned in previous chapters, the International Green Construction Code (IgCC) provides a comprehensive set of requirements intended to reduce the negative impact of buildings on the natural environment. Chapter 9 of the IgCC contains requirements to facilitate the pre- and post-occupancy commissioning, operation, and maintenance of buildings and to ensure the education of building owners and maintenance personnel with regard to best-operation and management practices. All of the provisions in IgCC Table 903.1 are mandatory except where they are indicated to be project electives. Project electives are selected from IgCC Table 303.1 by the owner or designer and may vary from project to project. The total number of project electives required to be implemented on each project is determined by the jurisdiction in Table 302.1. Where selected as a jurisdictional requirement in Table 302.1, IgCC Section 904.1.1 requires that reports verifying that the building is operating at required levels of performance be submitted to the code official, at intervals determined by the code official, throughout the building life. See Table 10-6 for a summary of the IgCC's commissioning, operation and maintenance requirements.

Table **10-6**
IgCC Commissioning, Operation, and Maintenance (Public Version 2.0)

Section 903: Commissioning

903.1.1 Pre-occupancy report requirement. The approved commissioning agency shall keep records of the commissioning required by Table 903.1. The approved agency shall furnish commissioning reports to the owner and the registered design professional in responsible charge and, upon request, to the code official. Reports shall indicate that work was or was not completed in conformance to approved construction documents. Discrepancies shall be brought to the immediate attention of the contractor for correction. Where discrepancies are not corrected, they shall be brought to the attention of the owner, code official and to the registered design professional in responsible charge prior to the completion of that phase of the work. Prior to the issuance of a Certificate of Occupancy, a final commissioning report shall be submitted to and accepted by the code official.

903.1.2 Post occupancy report requirement. Post-occupancy commissioning shall occur as specified in the applicable sections of this code. A post-occupancy commissioning report shall be provided to the owner within 30 months after the Certificate of Occupancy is issued for the project and shall be made available to the code official upon request.

Section 904: Building Operations, Maintenance and Owner Education

904.1 General. The operations and maintenance and building owner education documents shall be in accordance with Sections 904.3 and 904.4 and submitted to the owner prior to the issuance of the Certificate of Occupancy. Record documents shall be in accordance with Section 904.2. The building owner shall file a letter with the code official certifying the receipt of record documents and building owner education, operations and maintenance documents. At least one copy of these materials shall be in the possession of the owner and at least one additional copy shall remain with the building throughout the life of the structure.

904.2 Record documents. The cover sheet of the project record documents shall clearly indicate that at least one copy of the materials shall be in the possession of the owner and at least one additional copy shall remain with the building throughout the life of the structure. Record documents shall include all of the following:

1. Copies of the approved construction documents, including plans and specifications.

2. As-built plans and specifications indicating the actual locations of piping, ductwork, valves, controls, equipment, access panels, lighting and other similar components where they are concealed or are installed in locations other than those indicated on the approved construction documents.

A copy of the Certificate of Occupancy.

904.3 Building operations and maintenance documents. The building operations and maintenance documents shall consist of manufacturer's specifications and recommendations, programming procedures and data points, narratives, and other means of illustrating to the owner how the building, site and systems are intended to be maintained and operated. Operations and maintenance documents shall include items as specified in this code section.

904.4 Building owner education manual. The owner shall cause to be assembled an informational document on the building, site or structure and systems and sustainable features that are covered by this code and included in the building. Such information shall be educational in nature and sufficient for future tenants, owners and operators of the building, building site, structure and systems to understand the basic purpose and basis for these systems and features and how they are to be maintained for continued performance. The education documents shall consist of a statement of performance goals or requirements and a narrative illustrating the reasoning behind the building's site, features, and systems design. One copy of the owner education manual shall be in the possession of the owner and one additional copy shall remain with the building throughout the life of the structure. Where a whole building life cycle assessment is performed in accordance with Section 304, the data and final report shall be included in the owner education manual.

ACHIEVING COST-EFFECTIVE OPERATIONAL PERFORMANCE

Complex building systems and controls are increasingly becoming the new building norm. They require an increasing level of training and skill sets for building operators and maintenance staff. Whereas some owners invest wisely in their buildings for lowest life-cycle costs, optimal asset value, and reduced environmental impacts, others cut corners and costs by deferring maintenance and generally allowing building systems and assets to degrade. Building commissioning provides the foundation for record documents, building operations, maintenance requirements, and building owner/operator education. To ensure long-term viability of building performance, a building should engage in a re-commissioning every 5 to 7 years. This is the only way to achieve cost-effective operational performance while providing for a healthy and productive indoor environment.

11 | A LOOK INTO THE FUTURE

Green building has a place in our future. There is no question about it. The success and likely continued success of the human species and our presence on planet earth make that an inevitable eventuality. Our continued population growth, expanding environmental footprint, and rate of consumption, especially in developing economies, dictate it. The question is not whether green building has a place in our future, but at what pace it will become a baseline practice.

In the not too distant future, the building industry will expect green building codes and standards to be the new baseline. Green codes and standards will soon be considered to be as important a fixture of modern construction regulations as fire, building, plumbing, and other existing baseline codes and standards are today. In the short term, voluntary use of green and sustainable codes, standards, and rating systems has dramatically increased, at least as a percentage of new construction, in the current, admittedly weak, economy. Greater acceptance and experience with green building methodologies will lead to eventual mandatory adoptions. As more and more voluntary users realize the benefits of green building, fears of the unknown will subside and integrated project delivery methods will prevail. This, in turn, will also push more green and sustainable provisions into the mandatory baseline codes.

Green codes, standards, and rating systems are gaining prominence around the globe in both developed and developing economies. In developing economies, where population growth and environmental footprint continue to expand at rates that put obvious and real pressures on natural resources and the environment, green codes, standards, and rating systems are sometimes considered to be more important than baseline codes. In fact, many developing economies have no baseline codes in effect whatsoever. This is a major problem, as green codes and standards are not, and were never intended to be, stand-alone documents. All require a foundation of baseline life and safety codes in order to truly produce green buildings.

Does the availability of green codes and standards mean that rating systems will cease to exist? Not necessarily. Rating systems do an effective job of encouraging the consideration and weighting the benefit of many green practices that are often difficult or impossible to mandate. Even the IgCC, which incorporates project electives that are similar to the elective provisions in rating systems, does not do as effective a job of weighting and comparing the environmental benefits of green practices as rating systems do. Tools that allow owners and designers to see that one practice is more environmentally beneficial than another can drive intelligent, sustainable choices during the design process. On the other hand, green codes and standards that are written in mandatory language, such as the IgCC and ASHRAE 189.1, ensure that, where it is reasonable and beneficial, green practices are always implemented and will not be circumvented.

Rating systems have catapulted green and sustainable concepts tinto the mainstream building market and have unquestionably improved overall building performance. Although rating systems may have room to grow as assessment tools, they are sure to continue to improve and are likely to reinvent themselves in many ways, just as green codes and standards will be updated and improved through the code development cycle. Rating systems were the drivers that spurred the creation of green codes and standards, and they will continue to spur codes and standards to greater heights.

The question of what is good enough and how far the levels of performance must be raised in order to produce a built environment that is truly sustainable will continue to be asked. There are likely to be many solutions, rather than one. Rating systems, as well as the IgCC's project electives, are stretch tools that recognize performance beyond code minimums. Baseline codes do not accomplish this, nor were they intended to. Insurance companies, however, often give discounts for construction that exceeds code in the areas of durability and health, including many best practices that are addressed in green codes, standards, and rating systems. If there is a financial incentive to exceed codes, it follows that there are likely to be real benefits to building owners as well, when they are willing to pay the premium.

As the minimum requirements of baseline codes and standards are developed, questions asked typically revolve around what is safe, enforceable, effective, affordable, reasonable, and technically feasible. However, it is also appropriate to recognize, and not necessarily mandate, higher performance in certain areas. This is what green codes, standards, and rating systems do. It is appropriate to recognize and encourage the implementation of more environmentally beneficial practices over those that are less environmentally beneficial. Rating systems, and the IgCC's project electives to a lesser degree, accomplish this, whereas baseline codes and standards do not.

Green and sustainable requirements will continue to evolve from leading-edge green rating systems into mainstream minimum green codes and standards. As the most widely used mainstream rating systems become better indicators of building performance, it is possible that they may integrate life-cycle assessment, environmental product data, and other green and sustainable concepts much more effectively. This may put life-cycle assessment, including whole-building life-cycle assessment, into perspective and make it a much more understandable and useable tool. And as some requirements migrate from green codes and standards into baseline codes, green codes and standards may increasingly become more focused on building performance in all areas that are beyond code minimums.

As stated earlier, none of the green codes, standards, and rating systems currently available is capable of producing a built environment that is truly sustainable. We still are in desperate need of new technologies and methodologies that are capable of reducing the negative impacts of buildings on the environment, as well as those that may be capable of reversing the effects of environmentally damaging dated technologies and methods of construction. Green codes, standards, and rating systems will continue to push, pull, and leapfrog each other into the future and will incorporate new technologies and methodologies as they are developed. Someday green codes, standards, and rating systems will produce a built environment that is truly sustainable, has few negative impacts on ecosystems and natural resources, and may even restoratively give back more than it receives from the environment. But we are not anywhere close to being there yet.

As we move forward, building officials will become caretakers of the environment, as will design professionals, contractors, manufacturers, building owners, and the general public. We all must, and are beginning to, become increasingly aware of our effects on each other and our planet.

One of the few things we can be certain of is continual change. Environmental conditions, human needs and expectations, and technology will continue to evolve and change. In response, green codes, standards, and ratings systems, as well as the baseline codes that form their foundation, will continue to evolve as well. We still have much to learn and discover, and we will continually be required to grapple with how to apply that new knowledge in constructive, feasible, and sustainable ways. Not to do so would have too many unintended consequences, including many that we do not even have the ability to foresee.

GLOSSARY

A

acoustics   The production, transmission, and effect of sound.

agrifiber   Any fibrous material generated from bio-based products.

artificial lighting   Lighting achieved through anything other than sunlight.

atmospheric quality   The quality of surrounding air and measurement of pollutants.

B

bio-based material   A material or product other than food that is composed of or created from biological or renewable materials, such as bamboo, wood, or cotton.

biodegradable/compostable   A material that will break down into organic matter when exposed to certain conditions, reducing the load on landfills and benefiting soil.

biodiversity   Variation of life within an ecosystem.

biomagnification   The tendency of pollutants to increase in concentration as it moves through the food chain.

biomonitoring   The monitoring of substances in blood, urine, breast milk, hair, organs, and tissues.

biophilia   The human need to be connected to nature and living things.

building envelope   A separator between interior and exterior building environments or "skin" to help maintain indoor environment and control climate.

building integrated photovoltaics (BIPV)   Photovoltaic materials that are used to replace conventional building materials such as the roof or skylights.

building operating plan   Defines the conditions required by building management and occupants for successful operations of the building including desired temperature, humidity levels, fresh air ventilation, lighting levels and time-of day schedules.

building-related illness   Any recognizable illness that comes from the building, such as Legionnaire's disease.

building service life   The period of time a building or its parts are expected to function without major repair.

C

carbon cycle   The continuous exchange and movement of carbon between atmosphere, land, and ocean, including its presence in all living things.

carbon-neutral building   Buildings designed with the intent to reduce carbon emissions through increased efficiency in heating, cooling, and lighting. This includes reducing energy use and consuming carbon-free (non fossil fuel) energy sources.

carbon-neutral design   Design intended to reduce carbon dioxide (CO_2) emissions. Carbon dioxide is a greenhouse gas.

chemical body burden   Sometimes referred to as "total chemical load". The capacity of our bodies to absorb chemicals from the environment before immune systems are compromised.

closed-loop building materials   Materials that are recovered from other structures and infinitely recyclable.

commissioning report   Identifies problems that are likely to emerge or require particular attention on an on-going basis.

compact development   The ratio of the total building area to total site area. Increased density and mixed use buildings allows for increased walkability and a reduced dependence on transportation services.

compact fluorescent light (CFL) Energy-efficient lighting that creates light using much lower wattage than traditional incandescent lighting.

construction management-at-risk (negotiated work) A project method where the owner contracts separately with the design team and the contractor or construction manager, who works on the owner's behalf. Early in the design process, the construction manager is required to guarantee that the total construction costs will not exceed a maximum price.

construction waste management Redirecting construction waste so that it can be recovered and reused instead of disposed of in landfills and incineration facilities.

cradle to cradle The life cycle assessment from original source ("cradle") to use and repurposing for reuse ("cradle").

cradle to grave The life cycle assessment from original source ("cradle") to use and disposal ("grave").

D

daylighting Placing windows or other openings and reflective surfaces so that natural daylight provides effective internal lighting.

deconstruction The process of taking a building or structure apart piece-by-piece with the intent of repurposing as many of the materials or components as possible.

demand-controlled ventilation (DCV) Ventilation in which the intake of outside air is varied based on the amount of carbon dioxide or occupancy in a space.

design-bid-build (hard bid) A delivery method where the design team is selected by the owner and works on the owner's behalf to produce construction documents. General contractors bid on the project and in turn select subcontractors based on competitive bidding.

design-build A project delivery method in which one entity (design/builder) executes a single contract with the owner to provide design and construction services. Integrated Project Delivery principles can be applied to a variety of contractual arrangements.

E

ecological footprint A measurement of Earth's biological ad ecological ability to absorb human waste and demand.

ecological site planning Planning that emphasizes and protects natural resources and the biodiversity of surrounding land and wildlife.

ecology The study of the relationship between living things, including people and animals, and their environment.

ecosystem A set of organisms living in an area, their physical environment, and the interactions between them.

ecosystem services The conditions and processes through which natural ecosystems and species sustain human life. This includes both goods (such as food, oil, timber) and functions that support life, such as waste reduction and cleaning.

embodied energy The sum of energy input or work needed to make a product. Products with greater embodied energy usually have a higher environmental impact due to emissions and greenhouse gases associated with energy consumption. Recycled materials have lower embodied energy since they are not manufactured from raw materials.

embodied energy of water The energy needed to source, treat, and distribute municipal water.

emission The act of being released.

emissivity The measure of how well a surface radiates heat back out into space.

energy management systems (EMS) Software-based programs that monitor and control energy-consuming systems to ensure that the building operates efficiently and effectively.

envelope-dominated building Large surface area to volume ratio and small internal heat sources, and therefore greatly affected by climate (i.e. residences, small commercial buildings).

environmentally sensitive land An area that that contains rare natural features that need protection.

F

fenestration Window design and placement.

G

graywater reuse   Recycling water from sources like showers/bathtubs, sinks, and clothes washers (i.e. has not come into contact with wastewater) for reuse in irrigation, toilets, exterior washing, and other nonpotable uses.

green building   Building that consciously aims to reduce the negative impact of buildings on the environment.

green building rating systems   Rating systems that provide checklists and tools to measure the environmental impact and sustainability of a building. These systems often reference building codes and standards to measure building performance.

greenfield site   An area that has not been developed previously and includes undisturbed ecosystems and productive land such as forests and farmland.

greenhouse gases (GHG)   Gases that trap heat below the earth's atmosphere as the glass of a greenhouse traps heat produced by the sun.

greenwashing   The act of misleading consumers regarding the environmental benefits of a product or service or the environmental practices of a company.

H

heat island effect   The phenomenon whereby heat-absorbing building release heat absorbed from sunlight into the surrounding atmosphere, thereby increasing the local temperature.

heat transfer   The heating and cooling of a building done by the addition and subtraction of heat, including conduction, convection, and radiation.

high-performance building   Uses whole-building design to create a building that performs substantially better than standard buildings in energy use and economic and environmental impact.

hydrologic cycle   The continuous movement of water on, above, and below the Earth's surface, including its transformation between states (liquid, vapor, solid).

I

indoor environmental quality (IEQ)   The quality of air in a building environment.

infill development   The process of developing vacant or under-used tracts within otherwise developed areas.

insulating effect of thermal mass   Thermal mass can store heat for passive heating and can be used as a heat sink for passive cooling thereby eliminating peak demand for air conditioning.

integrated building design   Design that integrates multiple disciplines and seemingly unrelated design aspects for greater benefits and results.

integrated design   A holistic design approach that requires design conditions and variables be considered as a unified whole to achieve the desired outcome.

integrated project delivery   A project delivery method that requires a close collaboration between the owner, architect, and the general contractor responsible for construction of the project, from early design through project handover.

internally-dominated building   Less affected by climate because of a small surface area to volume ratio and large internal heat gains such as equipment and lighting (i.e. auditoriums, large commercial buildings).

L

life cycle assessment (LCA)   An evaluation of the environmental impacts of a building system (e.g., use of resources and environmental consequences) throughout its life cycle.

life cycle cost (LCC)   All costs of acquiring, operating, maintaining, and disposing of a building.

light pollution   Excessive or obtrusive artificial light.

light-emitting diodes (LEDs)   Light source with lower energy consumption and longer lifetime than incandescent light sources.

M

multiple chemical sensitivity (MCS)   A growing percentage of the population that is extremely sensitive to very small quantities of contaminants.

N

net-zero energy building   A building that produces on-site energy from renewable resources that at least equal s what the building consumes from other energy sources.

new urbanism   Urban planning movement designed to encourage pedestrian- based and mixed-use characteristics that were typical in pre-automobile communities.

nitrogen cycle   The process in which nitrogen is converted biologically and non-biologically. Nitrogen makes up the majority of Earth's atmosphere, but its lack of availability in some ecosystems can affect key processes like decomposition.

O

outdoor-indoor transmission class (OITC) A standard used to indicate the rate of sound transmission between outdoor and indoor areas.

P

passive design   The design of a building's heating, cooling, lighting, and ventilation systems, relying on sunlight, wind, vegetation, and other naturally occurring resources on the building site.

peak load reduction   Reduced consumption at times of peak demand.

pervious surface   A highly porous surface that allows water to pass directly through, in turn reducing runoff and allowing water to move downward from the surface into the ground.

post-consumer recycled content   Materials used in a product that are recycled after consumer use.

pre-consumer (post-industrial) recycled content   Materials used in a product that are recycled from the manufacturing process.

preventive maintenance plan   Should include the manufacturer's recommendations for the ongoing operation of the base building systems.

pv (amorphous silicon) modules   Solar panels that are crystalline and thin film.

R

radon   A cancer-causing gas produced from the breakdown of uranium in soil, rock, and water, and present in the indoor air of many homes.

rainwater harvesting   A collection system that collects and stores rainwater for other uses, such as irrigation and nonpotable uses when permitted by local code.

recyclable   A product's ability to be recycled after its useful life. Consideration should be given to the ease and accessibility of deconstruction and cost-effectiveness.

reflectivity   The measure of how well a material bounces back radiation.

responsible site development   Site development that protect habitats, reduces heat island effect, minimizes light pollution, maximizes pervious surfaces, retains native vegetation, and manages on-site storm water through reuse and infiltration.

S

semi-volatile organic compound (SVOC)   An organic compound with a boiling point higher than water and which may vaporize when exposed to high temperatures. SVOCs dont become gaseous as readily as VOCs but are still found in indoor air.

sequence of operations   Defines operational states desired for different conditions including operations for full or partial loads, staging or cycling of equipment, water temperatures, air pressures and operating phases for warm-up, occupied or unoccupied conditions.

sick building syndrome (SBS)   Symptoms in occupants of unhealthy buildings, including headaches, irritated eyes, nausea, fatigue, dizziness, coughing, and throat irritation. These symptoms tend to go away or improve significantly when they leave the building.

sidelighting   Lighting produced on the side of buildings, such as side windows.

site hardscape   Non-porous, hard surface that increase stormwater runoff, erosion, groundwater pollutants, and summer heat sinks.

site inventory   Identification of existing features of a site, including topography, hydrology, soils, and ecological features.

smart growth   A planning strategy to help communities develop and grow in a way that supports environmental conservation, economic development, and community vitality. Smart growth is transit and pedestrian-oriented;

concentrates development with a mix of housing, commercial, and retail uses; protects open and environmentally sensitive space; and recognizes the connection between development and quality of life.

solar reflectance index (SRI)   The measure of a material's ability to reject solar heat.

sound transmission class (STC)   A standard used to indicate how well a building partition reduces airborne sound.

stormwater management   Managing water that originates during storms and other events so that the volume and timing of the runoff does not affect water levels and supplies, and to mitigate or reduce the effect of contaminants carried by the runoff.

sustainable design   Design that uses healthy and renewable materials; efficient resources and energy; and is sensitive to economical, social, and ecological land-use and future sustainability with minimal impact on the natural environment.

sustainable development   Development that serves future as well as current needs. This includes favoring construction processes that restore the environment; recycle/reuse building materials; encourage self-reliance in communities to reduce transportation and energy use; and use materials that are naturally suited to the surrounding environment.

sustainable forestry   Forests with managed resources to meet product needs while maintaining long-term health and ecosystems.

systems narrative   A summary description of building systems including heating, cooling, ventilation, water heating, humidification and/or dehumidification, lighting and associated control systems.

T

toplighting   Lighting produced overhead, such as skylights.

transect planning   A system defined by a transect, or series of zones that transition from less density at the edge to high density in the town center.

V

variable-air volume (VAV)   A type of heating, ventilation, and/or air conditioning system that is an efficient alternative to constant-volume reheat systems.

vernacular buildings   Buildings constructed from local materials defined by the geology, ecology and climate of the region. These structures are highly practical, energy efficient, and blend with the landscape.

volatile organic compound (VOC)   Gases emitted from certain solids or liquids.

INDEX